Global IT Outsourcing
Software Development across Borders

This book offers key insights into how to manage software development across international boundaries. It is based on a series of case studies looking at the relationships between firms from North America, the UK, Japan and Korea and Indian software houses. In these case studies, which have typically been compiled over a 3–4-year timespan, the authors analyse the multi-faceted challenges encountered in managing these Global Software Alliances (GSAs). These challenges range from the conflicts that managers face when dealing with distance to the tensions of transferring knowledge across time and space, to issues in trying to establish universal standards in a context of constant change and the problems of identity that developers and clients experience in having to deal with different organizations and countries. Throughout the book, the authors draw on their extensive research and experience to offer constructive advice on how to manage GSAs more effectively.

Sundeep Sahay is a Professor in Informatics at the University of Oslo, Norway. After completing his doctoral studies at Florida International University, he has held research and teaching positions at the Universities of Cambridge and Salford in the UK, and at the University of Alberta in Canada.

Brian Nicholson is a lecturer in information systems at the University of Manchester, UK. After completing doctoral studies at the University of Salford, UK, his research interests have focused on the complexities of software development and software outsourcing between UK and Indian companies.

S. Krishna is a professor at the Indian Institute of Management, Bangalore. His research interests concern global software work (GSW) arrangements. He holds a PhD in software engineering and chairs IIM's software enterprise management programme, focusing on research and management education in partnership with local software industry.

Global IT Outsourcing

Software Development across Borders

Sundeep Sahay, University of Oslo, Norway,
Brian Nicholson, University of Manchester, UK
and S. Krishna, Indian Institute of Management, Bangalore, India

CAMBRIDGE
UNIVERSITY PRESS

PUBLISHED BY THE PRESS SYNDICATE OF THE UNIVERSITY OF CAMBRIDGE
The Pitt Building, Trumpington Street, Cambridge, United Kingdom

CAMBRIDGE UNIVERSITY PRESS
The Edinburgh Building, Cambridge CB2 2RU, UK
40 West 20th Street, New York, NY 10011–4211, USA
477 Williamstown Road, Port Melbourne, VIC 3207, Australia
Ruiz de Alarcón 13, 28014 Madrid, Spain
Dock House, The Waterfront, Cape Town 8001, South Africa

http://www.cambridge.org

First published 2003

Printed in the United Kingdom at the University Press, Cambridge

Typefaces Minion 10.5/13 pt. and Helvetica Neue   *System* LaTeX 2$_\varepsilon$ [TB]

A catalogue record for this book is available from the British Library

Library of Congress Cataloguing in Publication data

Sahay, Sundeep.
    Global IT outsourcing : software development across borders / Sundeep Sahay, Brian Nicholson, S. Krishna.
        p.   cm.
    Includes bibliographical references and index.
    ISBN 0-521-81604-1
    1. Computer software industry–Subcontracting.   2. Computer software–Development–Management.
    3. Information technology–Management.   4. Strategic alliances (Business)   5. Globalization–Economic
    aspects.   I. Title: Global information technology outsourcing : software development across borders.
    II. Nicholson, Brian, 1967–   III. Krishna, S.   IV. Title.
HD9696.63.A2S24 2003
005′.068′7–dc21   2003040949

ISBN 0 521 81604 1 hardback

Contents

1 Introducing the phenomenon of global software work   1

2 Globalization and global software work   27

3 GlobTel's GSA programme in India   51

Figures

Tables

Boxes

Foreword

Software is a key element in the increasing use of information and communication technologies in contemporary society, and thus its production and use are of major importance. Outsourcing of software production has been common for many years, but since the late 1980s this has increasingly occurred across national and cultural borders, a phenomenon which is known as 'global software outsourcing'. Cost is a major driver of this, with production normally located in countries with relatively low wage levels, but outsourcing organizations recognize that effective relationships with their software suppliers must be developed and maintained if full benefits are to be realized. This has resulted in various forms of collaborative arrangements, which can be labelled as 'Global Software Alliances' (GSAs).

This book provides rich empirical data on Global Software Work (GSW) and associated organizational alliances. The material is derived from the extensive fieldwork carried out by the authors over a number of years, with special emphasis on outsourcers in Canada, the UK and Japan, and companies in India as the software producers. India is a major success story in this area, with quite exceptional growth rates of its software export sector since the 1990s. However, as the book so vividly illustrates, this has not been achieved without lengthy and sometimes painful learning processes on the part of those involved on both sides of the outsourcing alliances. The longitudinal nature of the fieldwork carried out by the authors enabled them to trace and analyse such processes over several years.

The book can be read at one level, therefore, as a set of 'war stories' of the shifting objectives, personnel, relationships and outcomes of the various case studies. However, at another level, the book aims to connect these stories with the broader debate on globalization. The book argues that GSAs can be conceptualized as both a model *of* globalization and a model *for* globalization. In other words, that GSW, in its arrangement and conduct, both reflects globalization phenomena, such as the increasing interconnection of the world across time and space, and is itself one of the contributors to how globalization evolves, since all parties to it are affected by the ongoing process.

The authors also theorize their empirical work through an interesting set of six micro-level themes derived from a combination of their own experiences and aspects of the literature on software outsourcing. One theme, for example, examines tensions between the unbounded global space within which it is feasible in principle to conduct

GSW, and the boundedness of the individual software developer's need to belong in a particular place, with all that this entails in terms of family and broader social relationships. Other themes, all of which are related, examine shifting identity, the complexities of knowledge sharing, the limitations and benefits of standardization, issues of power and control and the challenges of cross-cultural communication.

The book will be essential reading for academics and other researchers working in the area of software production and outsourcing, but it will also be of interest to the wider community of scholars concerned with the role of information and communication technologies in the contemporary world, and in particular those trying to understand the phenomena known as 'globalization'. Thoughtful practitioners will also find much of value here. The book does not contain a set of prescriptions on 'how to do it' in all contexts, since the authors would argue that action needs to be context-specific. Nevertheless, they do offer sets of questions which the practitioner will wish to try to answer in their own specific context, based around the areas of the management of knowledge, people, communication and relationships within the overarching concept of managing ethically. The book aims to contribute to both theory and practice in the area of global software work, outsourcing and alliances, and I believe that it will achieve success in this endeavour.

Geoff Walsham

Acknowledgements

Six of the chapters in the book (4, 5, 6, 7, 8 and 9) are based on empirical material that has been written up in earlier papers (or are currently under review). The theoretical lenses applied for the analysis of the empirical material, content and conceptual framework are significantly different from those used in the earlier papers.

Chapter 4 draws upon the empirical material presented in S. Sahay, The challenge of standardization in global software alliances. This paper has been accepted for publication by the *Scandinavian Journal of Information Systems*.

Chapter 5 draws upon the empirical material presented in S. Krishna and S. Sahay, Evolution of global software outsourcing relationship and transformations in identity, Working Paper, Indian Institute of Management, Bangalore and University of Oslo, 2002 (the paper is currently under review in an international IS journal).

Chapter 6 draws upon the empirical material presented in S. Sahay and S. Krishna, An empirical investigation and a dialectical analysis of a global software outsourcing arrangement, Working Paper, University of Oslo, 2002 (the paper is currently under review in an international IS journal).

Chapter 7 draws upon the empirical material presented in B. Nicholson, S. Sahay and S. Krishna, Work practices and local improvisations within global software teams: a case study of a UK subsidiary in India, Proceedings of the IFIP 9.4 Conference on Socio-Economic Impacts of Computers in Developing Countries, Cape Town, 24–26 May.

Chapter 8 draws upon the empirical material presented in B. Nicholson and S. Sahay, The political and cultural implications of the globalization of software development: case experience from UK and India, *Information and Organisation*, 11, 1, 2001, 25–44.

Chapter 9 draws upon the empirical material presented in S. Krishna and S. Sahay, GSO experiences in Korea and Japan: some preliminary investigations, Final Report for the Project, The Context of Innovation of the Information Technology Industry, University of Pennsylvania Institute for the Advanced Study of India, New Delhi, August 2001.

For the conduct of different parts of the empirical research, we thank:
Professor Bob Hinings, University of Alberta, Edmonton, Canada
Dr Michael Barrett, University of Cambridge, Cambridge, UK

Dr Abhoy Ojha, Indian Institute of Management, Bangalore, India
Professor Geoff Walsham, University of Cambridge, Cambridge, UK.

For comments on earlier versions of the draft, we thank:
Dr Sarah Maxwell, Fordham University, New York, USA
Dr Eric Monteiro, NTNU, Trondheim, Norway
Dr Chris Westrup, University of Manchester, UK
Dr Ole Hanseth, University of Oslo, Oslo, Norway
Dr Susan Scott, London School of Economics, London, UK
Dr Sudi Sharifi, University of Salford, Salford, UK
Dr Erica Wagner, Cornell University, Ithaca, USA
Dr Mark Thompson, University of Cambridge, Cambridge, UK
Jayant Sahay, New Delhi, India
Shalini Sinha, New Delhi, India
Jonas Båfjord Holten, University of Oslo, Oslo, Norway

For institutions who have supported different parts of the research, we thank:
University of Alberta, Edmonton, Canada
University of Oslo, Oslo, Norway
Indian Institute of Management, Bangalore, India
University of Manchester, Manchester, UK
University of Salford, Salford, UK
European Institute, London School of Economics, London, UK
University of Pennsylvania Institute for the Advanced Study of India, New Delhi, India.

For firms that have been the sites for the different case studies in the empirical research, we thank (pseudonyms are used to preserve anonymity):
GlobTel, North America
Gowing, UK
Sierra, UK
Eron, India
MCI, India
ComSoft, India
Witech, India.

Abbreviations

ACCR	American Chamber of Commerce in Russia
ACM	Association of Computing Manufacturers
ASP	application service provider
BCS	British Computer Society
BPO	business process outsourcing
CCTA	Central Computer and Telecommunications Agency (from 1 April 2001 an integral part of the UK Office of Government Commerce)
CEO	chief executive officer
CMM	Capability Maturity Model
COO	Chief Operating Officer
DSP	digital switching product
DVD	digital versatile disc
EDA	electronic design automation
ERP	enterprise resource planning systems
EU	European Union
FTP	file transfer protocol
GIS	general information sessions
GRDG	Global R&D Group
GSA	global software alliance
GSODC	GlobTel Software Overseas Development Software Centre
GSW	global software work
HR	human resources
HRM	human resources management
ICT	information and communication technologies
IEEE	Institution of Electrical and Electronic Engineers
IIT	Indian Institute of Technology
IP	intellectual property
IPP	intellectual property protection
IPR	intellectual property rights
IS	information systems
ISO	International Standards Organization
IT	information technology

JV	joint venture
KPI	key performance indicator
MD	managing director
MNC	multinational corporation
NASDAQ	National Association of Securities Dealers Automated Quotation System (US)
NASSCOM	National Association of Software and Service Companies (New Delhi)
NGO	non-governmental organization
OECD	Organization for Economic Cooperation and Development
PCMM	People Capability Maturity Model
PDD	Performance Dimensions Dictionary
PR	public relations
R&D	research and development
SSADM	structured systems analysis and design methodology
UK	United Kingdom
UN	United Nations
USA	United States of America
WTO	World Trade Organization

1 Introducing the phenomenon of global software work

1.1 Introduction

It is 9 a.m. Monday morning and Peter Kelly, Managing Director of Academy Information Systems in Trowbridge, UK, has just arrived at his desk. He sits down to examine the progress on the latest release of Academy software for housing benefits. For three years this development has been outsourced to Mastek, an Indian software company. Kelly consults the Mastek website relating to the project and the 'dashboard' shows relevant indicators of quality, utilization, efficiency and schedule. Subsequently, he meets Sanjay, Mastek's project manager, who updates him on the progress verbally. Part of Academy's project team has arrived for work six hours before Kelly and has already made progress on several programming specifications given to them the previous evening. This is because the majority of the project team live and work in a different time zone, country and culture at Mastek's India development centre in Mumbai. Around lunchtime in the UK, before the Mumbai part of the project team leaves for home, they transfer the completed code to the server in Academy's Trowbridge office. The UK-based Mastek and Academy staff then have time for testing the completed code before incorporation into the beta release of the application. They can then prepare detailed specifications for the India-based team that they will pick up electronically in the Mumbai morning.

This brief story is an insight into the day-to-day life of Global Software Work (GSW), which is the topic of this book. We define GSW as 'software work undertaken at geographically separated locations across national boundaries in a coordinated fashion involving real time or asynchronous interaction'. GSW can thus include work done across global borders through outsourcing, alliances, or subsidiary arrangements. GSW is still an unexplored form of work and is enabled through organizational forms quite distinctive from traditional global arrangements as typified by large multinational corporations. Unlike manufacturing activities and professional services such as consultancies that have been studied in the past, software development in global settings remains empirically largely unexamined. Software development is a knowledge-intensive activity, and typifies work in the 'knowledge' or 'network' society. An analysis of such work in practice can provide interesting insights into the kind and extent of operations that can be effectively conducted in conditions of globalization. GSW takes place within an extremely dynamic and diverse global marketplace that is populated by organizations big and small from countries both developed and developing. The GSW arena is thus unique in that firms need not be fatally handicapped by existing size, and can potentially make an impact based on their knowledge competencies, ability to leverage technology and the cost advantages they offer. Diversity, complexity and uniqueness are thus inherent in GSW, making it an exciting and relatively unexplored domain of study. Analysis of GSW has implications for different disciplines concerned with such arrangements, including information systems, international management, computer supported collaborative work and organization theory. GSW arrangements are also of concern to policy makers responsible for economic growth and infrastructure develop-ment, particularly in developing countries such as India that has benefited greatly from an expanding export oriented software industry.

The aim of this book is to develop an empirically informed understanding of the process of Global Software Alliances (GSAs), the organizational arrangements that are established for the conduct of GSW. The evolution of GSAs are conceptualized within the context of globalization. We do this through the analysis of case studies that allow for an interrogation of various issues in the relationship from a variety of perspectives. Through inter-case comparisons, we seek to develop theoretical and managerial implications that can inform a better understanding about the conduct of the GSW phenomenon. The book can be read on two levels. First, and primarily, it can be treated as a study of global-ization, examining specific cases that are both a model *of* and a model *for* globalization. Secondly, our analysis will be of interest to managers and practitioners charged with the task of undertaking GSW who are prepared to go beyond simplistic 'how to'-type guides and methodologies. The strength of the approach is in the use of case studies to provide an in-depth analysis of particular issues, together with rigorous employment of theory that can help to develop both practical and theoretical implications.

We have structured the book in eleven chapters that may be read in a linear fashion from 'cover to cover' or as a reference and resource in the GSW area. Chapter 1 intro-duces the phenomenon of GSW. Chapter 2 provides an exposition of the theoretical underpinnings of the research approach and chapters 3–9 describe and analyse key themes within the detailed case studies of companies involved in GSW in India. These cases are the result of research undertaken during the period 1995–2000, involving some 200 interviews. Chapters 10 and 11 are concerned with the implications of our analysis at a theoretical and practical level, respectively.

The aim of this chapter is to present the GSW phenomenon in depth and set up the theoretical basis for the subsequent analysis. This phenomenon is shaped by three defining themes relating to the nature of the organizational form that enables such work, the kind of work that is conducted and the complex global trends within which such work is carried out. A detailed discussion of these three themes now follows.

1.2 Organizational forms and GSW

Globalization is a key characteristic of change in many domains at the turn of the twenty-first century. The most visible aspects of globalization have been a bewildering collage of transformations – increasing religious fundamentalism coexisting with greater secular human concern, development of centres of advanced technology amid regions of poverty and interconnectedness of systems and regions in ways that did not exist before. International business environments and organizational forms are being significantly reshaped as part of a new scenario that has variously been labelled as the 'new economy', the 'digital economy,' the 'network society', or the 'information age'. In these new environments, changes are especially visible in the kind of *organizational forms* being adopted to enable global work. A distinctive and defining aspect of these new forms is the manner in which *space and time* have become the primary medium through which to rethink the nature of the organization (Friedland and Boden 1997). An example of one such new organizational form is the 'Global Software Alliance' (GSA), a term we use to describe the nature of organizations established to enable GSW.

A GSA can be conceptualized as a relatively long-term inter-organizational relationship established between the outsourcing organization (the outcomes) and the outsourced organization (the contractor) based in different countries to enable software development in both real time and asynchronous time. This development occurs primarily in shared electronic domains with developers being located in the physical premises of their respective organizations (referred to as 'offshore'). The opening vegnette demonstrates a case of offshore programming and emphasizes the additional dimensions of managing projects across distance, time, language and culture. Offshore arrangements contrast with earlier 'body-shopping' (see below) where programmers from the outsourced firm carried out development while being physically located in the outsourcing organization (referred to as 'on-site'). Taking advantage of the increasing sophistication and capacity of telecommunication links and relatively lower labour costs in the outsourced organization, work in GSAs is done primarily in electronic spaces created through the use of information and communication technologies (ICTs) such as videoconference and email. While the physical travel of personnel between the vendor and contracting organizations can never be completely eliminated, the ongoing quest of both sides is to optimize costs by minimizing travel and finding the appropriate blend between on-site and offshore development. As GSAs seek to find synergies between remote and face-to-face work, time, space, organizational and national boundaries are recombined in novel ways where the experience of 'here' and 'now' loses its immediate spatio-temporal referents and becomes tied to and contingent on actors and actions at a distance.

Historically, the fortunes of firms in developing countries were seen as tied to the fortunes of those in the developed world. Our research into GSW provides some examples of firms in the developed world whose own fortunes are tied with equal potency to those in the developing world. Prior to GSW arrangements being possible, global work was primarily conducted by large organizations by virtue of their substantial direct investment transcending national borders. Based on their theory of a strategic mentality, Bartlett and Ghoshal (2000) categorized such firms as being *international, multinational, global* or *transnational*. At one end of the spectrum are international organizations that use their overseas operations in a marginal way, for example, simply to supply raw materials and marketing contacts to the parent company. At the other end, transnational organizations seek to integrate overseas operations more fundamentally by developing global efficiencies while also creating locally responsive approaches. In between, there is the multinational corporation (MNC) which takes a flexible approach by modifying its practices and products across countries. Managers of transnational organizations adopt a global outlook and seek to develop standardized approaches based on the assumption that there are more similarities than differences across countries. In centralized global companies, foreign units are dependent on headquarters for funds and expertise, but the transnational selectively centralizes some resources at home and some abroad in keeping with the need to respond flexibly to different issues. The transnational corporation is characterized not by structure alone but by formal organization, information systems (IS), culture and values.

Although the Bartlett and Ghoshal typology may still hold in the categorization of different kinds of software firms doing work globally, what is interesting is that these firms are quite different from those that have traditionally operated internationally. Size and ability of the firm to make large-scale investments on infrastructure are no longer terminally limiting factors in whether or not they can undertake GSW. Rapid upgrades in information and communication technologies (ICTs) have reduced the cost of communication and increased the scope of operations so that relatively small companies can potentially have business relationships and can address markets in different geographical domains. Some firms, particularly in such sectors as software, web development and other new media supported by networked and shared IT infrastructures, are capable of competing with larger companies in the global marketplace. Being an Indian or Russian firm is less of a perceived disadvantage and such firms are infact sometimes actively sought by larger ones by virtue of the knowledge capital they hold, the cost advantages they offer and the potential they provide to serve as a basis to access new markets. Along with large IT companies such as IBM and Microsoft there are many examples of firms who despite being small, are 'born global' and are capable of operating in a multitude of domains and countries (Saxenian 2001).

Saxenian argues that today new transportation and communication technologies permit even the smallest firms to build partnerships with foreign producers and tap overseas expertise, cost savings and markets. Start-ups in Silicon Valley today are often global actors 'from the day they begin their operations' (2001: 5). This multiplicity of networks in which these firms operate makes it difficult to categorize them on single dimensions of domains of work or countries of operations. They are better understood on their ability to develop and sustain *networks* that enable the flows of information, expertise, knowledge, and capital. Networks allow these firms to switch rapidly between local and global domains and build competence in different functional areas and markets. For example, Arrk, a small UK-based software house located in the University of Manchester Science Park employing only forty people, has the majority of its programmers in India and an international portfolio of customers. Cisco Systems has defined its core competence as product innovation, marketing/customer service and business relationship management. It delegates the rest, such as manufacturing assembly and product configuration, to its partners.

In operating these multiple networks, software firms deal not only with the strategic issues of whether or not and where they should globalize, but also with day-to-day operational issues including the creation of infrastructure, defining management processes and developing language and cultural understanding. Global projects have independent, autonomous links, and modules of work are distributed and coordinated using ICTs across wide physical and cultural distances. ICTs help both to intensify and redefine the nature of interactions across these different nodes which are not only confined to large organizations but also take place at the level of small firms and work teams. For reasons of geography and history, such as physical separation of different units and limited prior relationships of partners, these networks can comprise multiple short-lived global software teams (Carmel 1999). This is fundamentally different from firms composed of relatively autonomous units located in several countries as described by Bartlett and Ghoshal. However, the GSA relationship between the outsourcing and outsourced firms can take on different forms including joint ventures (JVs), vendor contract relationships and fully owned subsidiaries. New relationship models are also emerging: broker companies, for example, build databases of users and providers of outsourcing services and match firms based on predefined criteria. Some of these broker firms try to give more value than mere matching and provide project management services once the relationship is established. Another example is the 'hub' model where, for example, a Japanese firm may use its Singapore subsidiary through which to outsource to India. This model is used in an attempt to cost effectively and bridge some of the language, cultural and infrastructural gaps that would exist if work were carried out in India.

The organizational model adopted directly influences the pricing basis, that can vary from 'time and materials' to 'turnkey' or 'fixed price'. While in a time and material model development is priced on the programmers' time spent, in the other two cases, the basis is the estimated value of the whole project. The basis adopted has significant implications for intellectual property (IP) issues and the project control measures that need to be adopted. Where commitment in the relationship is not long term, and the aim is not to contract out new and core technologies, vendor contracts rather than JVs and subsidiary arrangements might be preferred. Relationships operate over different levels of a *trust continuum* (Heeks 1995) that is shaped by various considerations, including the length of the relationship, the kind of projects being done, the material investments made by both parties and the management capabilities to deal with the complexities of time, space and cultural distance. As the level of trust deepens, higher-end work can potentially be contracted out because of the increased level of confidence on both sides that work can be carried out effectively at a distance.

In summary, we have noted at least three distinctive aspects of GSAs:
- The manner in which different units of the network are *physically separated and electronically coordinated* across time, space and cultural boundaries.

- The ability to enter into GSAs is no longer restricted to large firms with the inherent capacity to make financial investments, but is also populated with small and innovative 'born global' firms driven by technology, ambition, intellectual capital and cost advantages.
- There is a central role for *ICTs*, for coordinating activities across different work units and for defining the content of work. Interdependent work requires the outsourcing and outsourced firms to be linked together by much higher bandwidth than that required for more stand-alone projects. While these ICTs help facilitate effective coordination and communication, they come with their own challenges related to access, compatibility, protocols and standards and issues of power and control.

  We build on these themes in the next section.

1.3 Nature of GSW

Software development and maintenance activities are the characteristics of processes of the 'new economy' involving programmers, software designers and analysts (collectively referred to as 'knowledge workers') engaged in designing, developing, testing and implementing software (referred to as 'knowledge work'). However, the nature of this knowledge is multi-faceted and continuously negotiated and contested by the various actors involved in the software development process (see for example, the case analysis of Sierra in chapter 7). GSW also reflects characteristics of other forms of global work in general where the focus is on developing standardization, productivity and efficiency. Ritzer (1996) labels such work as 'McDonaldization'. Based on an analysis of fast food restaurants, notably McDonald's, Ritzer develops a critique of current-day work practices and society as excessively concerned with institutions to *rationalize and control behaviour*. Drawing on Max Weber's views of rationalization, Ritzer identifies four dimensions of modern institutions:

- Efficiency
- Calculability
- Predictability
- Control.

As the case analysis of Witech in chapter 4 demonstrates, a constant quest in GSAs is to standardize and make efficient various aspects of infrastructure and work practices including, for example, defining the template in which project-related communication takes place. This quest for standardization and efficiency can also be viewed in the historical context of the software engineering tradition, and the continued attempt to impart structure and predictability to software development processes.

GSW involves the application of various kinds of *knowledge systems*, including programming languages, software development methodologies, project management techniques and the application domain. Different programming languages are used in software development, from the older FORTRAN and COBOL to the current Java and Visual Basic. Several hundred programming languages have been developed for use in both general-purpose and specialized domains. In the 1960s and 1970s, as technology of language compilers developed, large IT firms like IBM, Hewlett Packard and Univac formulated their own languages to support proprietary operating systems and system utilities. Users in other domains also developed their own languages – for example Nortel Networks, a large telecommunications firm, had software for their digital switches written in a proprietary language called Protel. A key technology for GSW came in the 1980s. Common standards increasingly emerged and C and then C++ (considered 'open' platforms) became widely used for system software development. Although the development of standards remains contested, developers preferred these open platforms as these did not restrict them to particular technologies, or to specific firms with their proprietary languages and products.

Although global work is not a new phenomenon, distributed software development work is relatively new and begs the empirical question: can approaches to global manufacturing (for example, car assembly plants) or global services (for example, consulting) be transferred seamlessly to software development work? As software work involves physically intangible artefacts whose value is derived from qualities such as efficiency of algorithms, 'look and feel' aspects of the user interface, richness of features and so on, this distinction from the production of material goods is useful. Software work has distinctive features, for example, in contrast to manufacturing where production and consumption take place in separate physical domains, services are generally distinguished by the *inseparability* of these functions. This is true of a range of different services from hotels and medical work to legal and accounting practices. However, these services are also starting to be outsourced offshore, as reflected in the growth of firms providing legal and medical transcription services and also those specializing in various transaction-processing functions like billing and ticketing.

Production and consumption are separable to a major degree in software work, where at each stage of the development, artefacts such as program code and documentation enable outputs to be specified and disembedded from the development domain to other use situations. However, information systems research has increasingly established that software design and development is never really 'finished', but involves an ongoing interaction and redefinition with the process of use (Bjerknes, Bratteteig and Espeseth 1991). Development and use of software can thus be quite distinct, linked together by various artefacts, and simultaneously be also intricately interconnected. Managing these *complex interdependencies* is a defining aspect of GSW.

Software may be regarded as a knowledge industry but is different from the traditionally accepted knowledge work of consulting in which many aspects rely fundamentally on the expertise of individuals, making it difficult to obtain economies of scale. Software work covers a range of activities including the development of algorithms and user interface designs that require creative talent of the highest order that cannot be scaled up in a mechanical fashion. Friedman (1989) points out that software work of this nature is continually being disciplined, formalized and made subject to managerial control, but that this is thwarted by factors such as rapid changes in technology and the associated lack of skills in these new domains. Other activities in the spectrum of software work include the work of call centres, data entry and medical and legal transcription that typically need a minimum level of English, typing skills and ability to use a word-processing program. Such work can easily be scaled up with a suitable work place and telecommunications infrastructure, wherever people with these minimal background skills are available in large numbers. In between these extremes, there is a range of activities that demands different degrees of knowledge and skills, and is amenable to varying degrees of scaling up. For example, while maintaining legacy software does not need creative talent of the highest order, it needs individuals who can, in a short span of time, learn new languages, understand the complex relationships in a large piece of software and sensitively operate in the use domain. The extent of separability and scaling, therefore, varies for different software tasks and is significantly shaped by the infrastructure in place, including the available bandwidth, the degree of sophistication of management processes and the prior experience of the partners.

In GSW, tasks at various stages of the software life-cycle may be separated and implemented at different geographic locations coordinated through the use of ICTs. Maintenance and testing were among the first tasks to be outsourced, while early life-cycle tasks such as design and user requirements analysis were considered more difficult to contract out as they required more intimate knowledge of the firm's work practices as compared to maintenance and testing. On the face of it, those types of technology oriented development appear better suited for outsourcing where specifications can be developed and given to an outside party to execute. However, design tasks become harder to undertake because they assume a close familiarity with the market and user preferences. Alternatively, in modular approaches, modules of the software are divided into independent modules and its development 'outsourced' to teams in different locations.

Intangibility, heterogeneity, mobility and scalability are features that differentiate software work from other services and also manufacturing activity. The mental or intellectual activity involved in software work is captured in a form not tangible in the literal sense of being touchable by a human hand but nevertheless is made perceptible through magnetic or optical readers and other devices. The heterogeneity of software work is often limited by the standardization of development processes, methodologies and programming languages. While new and innovative work involves heterogeneity at early stages of conceptualization and design, it requires less at the stages of testing and implementation. Standardization of processes is central to disembedding and fragmenting of software processes to make them amenable to GSW. Perishability, especially important in services like hotels, is not so in software since artefacts like software code and manuals provide mobility with the use of ICTs and enable the life of the software to endure over time.

Another distinctive aspect of software work is the variety of social and human issues that come into play in the phases of design, development, implementation and interpretation of its longer-term implications. Software work, when carried out in a global setting, magnifies these complexities as it involves relationships of people, teams, organizations and nations with different backgrounds, spoken languages and styles of working in conditions of temporal and spatial separation. Standardization, which is the key to coordinating distributed work, is extremely hard to implement because of the complexities of GSW. Whereas many firms in the manufacturing and services industry try to downplay national and cultural issues through standardization, managers of certain GSAs may capitalize on local ideosyncracies, strengths and creative energies (for example, the case of ComSoft described in chapter 5). While large MNCs are widely seen as weakening of local cultural values, smaller software firms (like ComSoft), in contrast, often attempt to reassert these identities in an effort distinctively to define themselves, drawing upon resources like national and cultural identities in the face of global competition. An ongoing challenge is how to find the appropriate blend between universal solutions and local particularities in a context inherently characterized by a multiplicity of networks.

Another key feature of GSW is the manner and speed in which its knowledge content is subject to change and radical readjustment and is characteristic of work in Castells' (1996) 'network society'. Founded fundamentally on technology and information, GSW involves a new form of 'informational capitalism' in which time, space and knowledge are key resources that entities try to dominate and standardize using various organizational forms. Changes, both technological and organizational, are the norm in GSW. The rapid uptake of Web-based systems by businesses has radically changed the skill sets (like Java) required for software development, for example. Changes are taking place at many levels, from the business models adopted to specific policies and procedures implemented to stem the attrition rate among developers. To deal with these rapid changes, firms need reflexively to monitor and modify their processes on an ongoing basis.

GSW, as we use the term, is broader than the traditional 'software outsourcing' that involved the purchase of goods or services previously obtained internally. Box 1.1 shows Apte's (1990) summary of the activities that were typically outsourced in the past. Prior to the early 1980s, the involvement of foreign companies in outsourced work of this nature was restricted mainly to data-processing or coding-type projects completed by a team of on-site foreign contract programmers. GSW now commonly involves the design and development of new products, support, special services, and whole-life-cycle projects involving different levels of complexity. While 'outsourcing' refers to work contracted out to third-party firms, typically located in the same country, GSW includes work done by subsidiaries and alliances that are necessarily located in a different country. GSW involves turning over to this third party or offshore subsidiary some or all the software development and maintenance tasks, ranging from simple data entry or programming to complete software design, development, data centre operations and full system integration. As the software component in hardware such as telephones, DVDs, cell phones and in cars and airplanes increases, the demand for outsourcing has multiplied (Box 1.1).

> **Box 1.1 Outsourced IT services**
>
> - Data-processing services: data entry, transaction processing, back-office clerical tasks
> - Contract programming
> - Facilities management: operation and support of data centres
> - System integration
> - Support operations: maintenance services and data recovery
> - Special services: training, hotline support

Source: Apte (1990).

GSAs allow for a range of possibilities both in terms of the kind of projects contracted out and the extent to which the different stages of the development life-cycle can be outsourced. Issues that influence these decisions are the strategic importance of the activity, the degree to which the requirements can be specified and the comparative cost of having the software developed in-house versus having it outsourced. Various arguments have been made for and against companies contracting out core projects: 'never outsource a problem, only a defined task' (Willcocks, Fitzgerald and Lacity 1996; Willcocks and Sauer 2000). Gurbaxani (1996) argues, to the contrary, that innovative projects can be outsourced with support of a strong contracting structure, presence of multiple vendors and a selective outsourcing strategy. In addition to this debate on what should be outsourced, there are also discussions on whether firms should go for 'total' or 'selective' outsourcing, ranging from small stand-alone projects completed through short-term employment of programmers, to projects where the third-party vendor is completely in charge of hardware, software and staff. Total outsourcing can involve design, implementation and maintenance of large projects, or even the support for whole pieces of legacy software or the porting of them to alternative platforms.

While the issues referred to above have been debated extensively in the information systems literature, they remain open empirical questions in the GSW domain. McFarlan (1995) has argued that 'highly structured' projects where processing, file structures and outputs are completely defined, are easy to outsource, as compared to those whose outputs are more open to the users' changing judgement on desirable features, for example, in business process re-engineering projects. Such projects, it is argued, are best done in-house in conditions of co-location and proximity. In addition, to reduce ambiguities in design, Kobitzch, Rombach and Feldmann (2001) argue that structured projects are amenable to an 'engineering' approach which makes it easier to scale up and achieve economies of scale that justify the investments required for establishing a GSW infrastructure. This is related to the maturity and compatibility of the structured processes in both the outsourcing vendor and client organizations. Our empirical work in Japan (see chapter 9) suggests that firms there may be more willing to operate in relatively unstructured projects as compared to North American firms, and this can potentially challenge the popular contention about the need for structure in remote work. While it is still early to say whether the Japanese approach will be successful, their experience will provide key learning for the practice of GSW.

In summary, we have pointed out four distinctive features of GSW:

- GSW does not reflect either a traditional manufacturing or service activity, and includes *elements of both.*
- GSW can take on varying levels of *sophistication and need for creative and intellectual inputs*, ranging from call centres to designing new technologies.
- The *scalability* of GSW varies with the nature of work and the life-cycle stage of the project.
- *Social and human issues* in GSWs are magnified as compared to traditional outsourcing because of the increased diversities of people, practices and technologies involved.

1.4 Global trends in GSW

What makes GSW particularly interesting is the diverse global network in which it takes place. This diversity is expected to increase in the future since different predictions on the size of the GSW market present a healthy picture of growth that no doubt will attract new producers. The demand for and supply of GSW has increased substantially since the early 1990s with continued demand from well-established users in the USA, UK, Australia and various Western European countries. Japan and Korea are fast emerging as customers and in some cases are also aspirant producers of various categories of GSW. According to the International Data Corporation (reported in Krill 2001), the USA is likely to continue to be the most significant user of GSW, with predicted increases in spending from $5.5 billion in 2000 to a rather optimistic $17.6 billion by 2005. At the time of writing, such predictions on the size and growth of GSW can be considered less than reliable because of the US recession and war in the Middle East. However, substantial growth of GSW throughout the 1990s was stimulated by such factors as demand for 'e-enablement' of enterprises and an international shortage in IT skills that forced firms to look offshore for resources. Other factors concern continued momentum towards IT outsourcing in countries such as Japan and Korea, in part due to the gradual erosion of the guaranteed lifelong employment structure. On the supply side, deregulation of markets in a number of developing countries, and a series of initiatives in countries such as China to address issues of English language and telecommunications infrastructure, have helped to position previously 'inactive' countries more prominently in the global marketplace.

Castells (1996) presents a view of the informational global economy as a network organised around *key metropolitan centres dispersed globally.* These centres comprise important 'nodes' for the structuring and management of intertwined activities at different levels, including firms, teams and individual actors. Castells' notion of a 'network' refers to both the physical (transport, computer systems and telecommunications) as well as the social (educational, expatriate networks, trade associations, personal contacts, etc.), in which economic, social, religious and even criminal activities need to be considered. The financial capitals of London, Tokyo and New York are good examples of major 'command and control' centres supported by various secondary nodes for the execution of various functions. Similarly, GSAs can be conceptualized as networks of users and providers of GSW services, infrastructure suppliers and a host of other actors. In GSW, global nodes situated around cities providing international market access to skills and knowledge structures serve global networks of firms. Telecommunications systems allow dispersion of these nodes around the globe, yet the nodes are themselves characterized by dispersion and concentration. The actors in these networks are connected by flows of various types of information, knowledge and capital, which help to shape the networks and also open up the possibilities of establishing new nodes. For example, Indian firms, through the experience and knowledge gained with North American companies, are actively establishing new markets in Japan and Korea. An implication of this networked perspective on GSAs is that it is inadequate to consider relationships only in one-to-one configurations (for example, an Indian and a North American firm), but to situate the analysis within a wider web of users and providers operating in the global marketplace.

In the GSW marketplace today, joint dominance exists between the major 'technopoles' of Ireland, India and Israel that are being challenged by a number of other emergent centres such as Russia, the Philippines and China. Within these countries, the cities of Dublin, Bangalore, Moscow, Manila and Beijing comprise major GSW centres that in turn are connected in global markets and social networks, and also help to create various secondary nodes internally. Within each country, the networking architecture reproduces itself in dispersed regional and local centres such as Mumbai, Hyderabad and Delhi in India and Novosibirsk, Moscow and St Petersburg in Russia. These 'networks of networks' constitute a process by which the production and consumption of GSW are connected through information flows. Global (e.g. the USA and India) and internal ('in-country' or 'in-city') networks contain specialization and fierce internal and external competition between cities in different regions and states. Mumbai, Bangalore and Hyderabad firms, while competing with each other, also vie for global contracts, bidding against firms based in Dublin and Shanghai.

We now discuss some of the global nodes and emergent centres in the GSW network, with an emphasis on the specialization strengths and weaknesses, with a view to describing the complexity of this network. We start with the 'big three' nodes, India, Ireland and Israel before considering three major emergent GSW centres of the Philippines, Russia and China. (Owing to space constraints, we are not able to discuss other emergent centres such as Singapore, Malaysia, Pakistan and Sri Lanka in Asia; Brazil and Chile in South America; Hungary, Romania and Ukraine in Eastern Europe; and Egypt in the Middle East.) We also discuss the trend towards 'nearshore' software development that involves Canada, Mexico, the Caribbean and Venezuela.

India

Many estimates of software exports consider India to be the leader in GSW, registering an average annual growth of software exports of more than 40 per cent during the 1990s. The trade association NASSCOM (www.nasscom.org) estimate that India's IT software and services exports is worth $7.7 billion in software for foreign clients and there are plans to expand this to $50 billion by 2008. Software services currently account for 10.5 per cent of India's total exports, making GSW one of the key engines for growth of the economy. India has come a long way from the situation in 1985 when Texas Instruments first saw potential and, in a pioneering and landmark move, established their subsidiary in Bangalore. Texas Instruments realized the limited capacity of the Indian government to provide infrastructure and obtained permission to establish their own satellite links and related infrastructure. The remarkable success of this centre to undertake leading-edge work inspired other (mostly North American) firms to establish facilities in India, particularly in Bangalore. A number of these firms (for example, Motorola) aimed to do leading-edge work, and their Bangalore lab attained the highest possible quality rating for software processes of Capability Maturity Model (CMM) Level 5; in 1992, only one other centre in the world (IBM) achieved a similar rating. Motorola's success opened the floodgates for software work in India, and other firms followed, a few entering through the route of JVs, many through wholly owned subsidiaries and contract staff both on-site and offshore. About 265 of the *Fortune 500* companies are now customers of the Indian IT industry and leading companies (e.g. Microsoft, SAP, Adobe and Quark) are setting up development centres. The success of companies like Motorola and Texas Instruments challenges the popular criticism that India has been a centre for only 'lower-end' work. There are currently more than 3,000 Indian software services exporters doing business in more than 100 countries; 25 companies account for 60 per cent of the sector's revenues and it is this vanguard that is spurring the global expansion and development of the IT services' value chain.

A majority of the initial software work in India was of the type derogatively called 'body-shopping' whereby the developers would go on-site for the length of the project. This trend has been steadily changing and although in 1990, about 95 per cent of the work was done onsite and 5 per cent in India, in 1995 two-thirds of all software services export earnings were created by on-site work, and currently about 70 per cent of a given project's development is done in India. Although development in India as opposed to on-site means lower staff costs, it also presents tremendous challenges in managing attrition and coordinating distributed work.

While Indian companies have come a long way in building expertise in project man-agement to deal with conditions of separation, they will in the future need to cope with further challenges arising from new competition (for example, China), and in develop-ing new markets (in Europe and East Asia), building expertise in new technologies (for example, in mobile telephony), upgrading infrastructure and articulating new models for pricing and profit sharing as compared to the 'time and material' approaches of the past. Reliance on the US market has made India's suppliers susceptible to reductions in IT spending and Indian companies have begun to explore opportunities in Europe.

Ireland

Ireland is often cited as the world's second largest software exporter after the USA with software exports ranging from an estimated US$4 billion in 1998 to US$8 billion in 2000 (Moore 2001). A large proportion of investment in Europe by US companies goes to Ireland, making them the second biggest exporter of software after the USA. They produce about 60 per cent of the packaged software sold in Europe. Ireland has the advantage of a strong technological infrastructure, EU membership, a sound technical education system, English-language competence, proximity and cultural similarity with the UK and the USA. Although arguably not as innovative or entrepreneurial as Israel, the Irish software industry turns out software products as well as a variety of support services. O'Riain (1997) traces the growth of the industry after 1973 when major MNCs were attracted by the Irish Industrial Development Authority's policies of financial incentives and significant investment in education and telecommunications. The late 1980s led to the arrival of sophisticated systems software companies such as Iona that were tempted by tax incentives and an attractive location to supply to the EU. The early 1990s saw European unification, that helped a booming Irish IT industry to grow at the rate of 20 per cent for most of the decade. Almost 80 per cent of the Irish software industry's output is exported.

By the end of 1998, an estimated 20,000 people were employed in the Irish ICT sector and total exports were valued at £4.2 billion. In 1999, there were some 570 indigenous software companies, 108 of which were foreign owned through arrangements such as subsidiaries. The major focus of the work was at the system level, including programming languages and tools for data management and data mining. Companies have been building on this core in order to create software applications specific to enterprises and industries. Most Irish companies are in software services and bespoke development with an emphasis on providing Internet and multimedia consulting. The main markets served are financial services, telecommunications, middleware, e-commerce and specific localization. O'Riain (1997) points out that unlike in India, Ireland's development has avoided relying on contract programming or 'body-shopping' and, instead, many large MNCs including Anderson Consulting, Intel, Digital, SAP, Sun Microsystems, Ericsson and Prudential Insurance have chosen to locate in Ireland.

The growth of the Irish software industry has been enabled by the ready availability of skilled staff, low corporate taxation, generous incentives, low operating costs and world-class infrastructure including telecommunications. The government has played a key role in providing financial support to companies to set up and expand overseas (Cochran 2001). The software sector has been made a strategic priority, and support has come in the form of legislation on security and copyright as well as funding for research and development (R&D). This has helped to give the industry a high-quality, low-risk image, and most of the firms have ISO accreditation or CMM of Level 2 or higher. The industry has been given a tremendous boost by the return of large numbers of technically trained expatriate Irish from the USA with a desire to contribute to the development of the country. The software industry thus is in a position of strength drawing from the state support, the strong inward investments by MNCs and a robust people base.

O'Riain (1997) identifies three future challenges to the industry:

- The industry's continued reliance on MNCs makes them potentially vulnerable to the risk of the companies deciding to shift their operations to other more attractive nodes in the wider GSW network.
- Irish companies need effective strategies to sustain their presence in the global economy in the face of a 'brain drain' of their educated people, especially the younger ones, primarily to the USA.
- The weak venture capital base and funding for R&D as compared to Israel, serves to restrict entrepreneurial development and the creation of new technologies.

Israel

Israel has emerged as a major global player in software exports, being one of the few countries able to seize the opportunities that globalization provided in terms of technology and knowledge, organizational forms, capital markets and specialized skills. Growth in the ICT sector started as early as in 1948, and Motorola set up their first subsidiary outside of the USA in 1964 (Ariav and Goodman 1994). Contributing to this early growth was the key influence of military-trained computing graduates who after completing their service entered the flourishing civilian computing sector. This trend still continues.

The military-inspired growth of the 1980s was further strengthened in the 1990s with the growth of high-tech 'clusters' involving start-up and venture capital firms strongly linked with the high-tech clusters of the US Silicon Valley and elsewhere (Teubal 2001). A number of new high-tech start-up companies opened in the late 1990s – about 2,500 firms in 1998 and 1,000 in 1999. In 2001, there were over 4,000 start-up firms and 120 organizations listed on NASDAQ. Venture capital from private, public and foreign sources has been the key in providing the impetus to growth in software exports. Today Israel has the second largest number of technology firms listed on NASDAQ after the USA.

The combined factors of public policy initiatives such as high R&D spending, a highly educated population, English-language ability, tax incentives, marketing support for software exports and a large expatriate Jewish population has facilitated strong links to markets abroad, especially the USA. The availability of high-quality telecommunications services offered by several competing providers has created a cheap and reliable infrastructure to support software work. Israel has a large proportion of technically skilled people, and connections with bankers and investors in the West. The return of foreign-educated Israelis has supported continuing technology transfer and created a demanding and high-quality consumer base that gives the industry its reputation for being of high 'battle tested' quality. The process of globalization has fostered in Israel a new kind of export goods – the sale of high-tech start-up companies to MNCs. Teubal (2001) calls this sale of technological assets rather than traditional merchandise and service exports one of Israel's most important export categories.

The centres of Tel-Aviv, Haifa and Jerusalem form the key software development areas, and between 1984 and 1992, the Israeli software industry tripled its sales and increased exports by 2,700 per cent. This trend has continued, reaching $1.3 billion in the first half of 2000, a 40 per cent jump from 1999 (IPR, 2000). There are about 300 software houses in Israel employing around 20,000 people. Nearly a third of Israel's software exports are sold to the USA, the remainder to Europe, an increase in part due to Israel's specialization in the Internet and communications sectors. Other areas of export include database management systems, application generators, computer centre operation, educational software and anti-virus protection. Israel also exhibits strength in niche areas of quality assurance of products and tools, security systems for the Internet and distance education. Major companies including Microsoft, IBM, Intel and National Semiconductor have Israeli subsidiaries (Ariav and Goodman 1994).

Israel shows the promise of becoming stronger and reaching an equal status with Ireland and India as a major provider of software. Israel's industry however differs fundamentally from India in that the key focus is on software products rather than on services. Compared to India, Israel has the advantage of being closer to the North American and European market. As a result of a service orientation, Indian firms have needed to familiarize themselves with customers while Israeli companies have chosen to compete internationally by developing technological assets that require less local interaction. Israel, like India, was affected by the 2001 slowdown of the US economy, and will need to redefine their future growth strategies. In addition, De Fontenay and Carmel (forthcoming) point out that the conflict with Palestine has affected growth of the industry, with foreign-customer concerns over safety and reliability in the event of increased violence. The current dependence on military-trained personnel to drive the civilian software industry may also pose a threat, since transfer of military-based technology may be less relevant in the future global scenario.

Russia

Although not one of the 'big three' software producers, Russia competes with India for offshore contracts from the USA and Europe. The McKinsey Global Institute estimates that Russia's offshore programming sector will grow at a rate of 50–60 per cent a year in the early twenty-first century. Many large firms including IBM, Nortel, Sun Microsystems, Boeing, Motorola, Intel, SAP and Microsoft have already started operations in Russia. A recent report (ACCR 2001) indicates that there are 5,000–8,000 programmers in Russia and annual revenue is between $60 and $100 million, reflecting a

40–60 per cent average annual growth rate. Of software exports, around 30 per cent are products and the remainder software services, typically offshore programming work (Heeks 1999; Lakaeva 2000).

Russia has a number of advantages including costs (salaries half of even Indian wages), a high-quality technical education and the third highest *per capita* number of scientists and engineers in the world. Many of these scientists had experience in nuclear, space, military and communications projects and moved into the software industry after the collapse of the Cold War. Mathematics and physics are strong areas in the skill base and Russian students are often winners in international programming contests. The cities of St Petersburg, Moscow, Vladivostok and Novosibirsk in Siberia are emerging as 'clusters' or 'silicon cities'. Proximity to markets in Western Europe and shared culture and history potentially reduce cross-cultural differences relative to India or China. Novosibirsk has special relevance for Germany owing to a large ethnic German population. This makes Russia an attractive potential partner for German companies as compared to Indian companies which have the handicap of the German language.

Many of the Russian offshore companies have 50–300 programmers and are partly or fully foreign owned. Smaller companies with 10–20 programmers rely more on links with friends and acquaintances to gain smaller contracts. Typical firms offer a list of services including Internet programming, Web design, Web server applications, database projects, system programming, real time and embedded systems, internationalization, translation and localization of software. The major end-users of these products and services include financial institutions, governments, educational institutions, industry and telecommunication Internet companies. Text recognition, anti-virus programs and the entertainment sector are other areas where export success has been achieved by Russian firms. High-level scientific work is also being done in Russia. Intel opened a subsidiary in Nizhy Novgorod in June 2000 to develop and support software for the next-generation Pentium processor.

While the future for Russia's software industry seems good, it needs to strengthen its institutional infrastructure by building something equivalent to India's NASSCOM. Russian companies also need to develop more sophistication in North American and European business practices and project management and to develop quality control processes that are in line with international standards. Growth has come despite the poor image of intellectual property protection (IPP) in Russian organizations. English is not as widely spoken as in Israel or India and the costs of bandwidth are higher than in other major offshore outsourcing countries. Other emerging vendor nations outside, but in the region of Russia, include Bulgaria, the Czech Republic, Hungary, Lithuania, Poland and Ukraine.

The Philippines

The Philippines is emerging as a key venue for offshore development, second to India in Asia, and is already a strong contender in a broad range of back-office services.

The Philippines' offshore software industry partly emerged as a result of the volcanic disaster of 1991 and the withdrawal of the US military in 1992, leaving behind a relatively reliable infrastructure that could still support a range of services. The Philippine government capitalized on this and developed trade around telecommunication and IT-enabled services as contrasted with India's development strategy in software services. Software exports for 2000 were estimated at US$200 million but it is not clear exactly what aspects this figure covers. There are some 30,000 Filipinos in the IT-related sector in several hundred firms, many of which are foreign owned (Hamlin 2001). The Philippines' telecommunication services include call centres and data processing as well as IT services such as applications development, Web design, animation, database design, networking and software. A survey in 2000 by META, a US research group, ranked the Philippines number one among forty-seven countries in the 'knowledge jobs' category. Software 'clusters' have been set up in Subic Bay and the Clark Special Economic Zone, with airports, telecommunications, housing complexes and tourist facilities. This has attracted back-office operations mainly from the USA – MNCs such as Barnes & Noble, Arthur Andersen and America Online.

The strengths of the Philippines include a good IT infrastructure especially in Manila and Clark IT parks, low labour costs (30–40 per cent less than in the USA) a highly literate (94 per cent) population, and a high level of English-language proficiency. A strong industry association in the IT and e-commerce Council helps to present a positive picture of the industry with comprehensive information stressing the quality of service and life in the Philippines. However, a record of political instability and a relatively poor general infrastructure still inhibit foreign investment. While the main telecommunications companies are expanding rapidly in the IT parks where the infrastructure is well developed, other parts of the country are still lacking in good-quality business accommodation, roads and support services. The geographical spread of the country, comprising 7,107 islands, makes it difficult to establish fixed telecommunication lines. Philippine law has also been slow to catch up with the new economy, a factor that may deter some MNCs from setting up operations because of IP fears.

China

China represents a major emerging supplier of software services especially after World Trade Organization (WTO) entry in November 2001. The Chinese software industry has grown at more than 20 per cent a year since the early 1990s, which is above the world average. The growth in 1999 was 30 per cent, to $2.16 billion, and future predictions are spectacular, on a par with India's success. China is perceived as a future threat to India in part owing to a relatively advanced user base of mobile phones and many more telephone connections and a vast pool of skilled human resources. The role of the Chinese government has been especially significant in attracting Chinese students in the USA to return home and establish new high-tech ventures. Saxenian (2001) quotes a survey that shows that about 18.8 per cent (around 160,000) of the Chinese students who studied in the USA between 1978 and 1998 returned to China to participate in these new ventures. This trend is significant as it supports the development of transnational networks of Chinese entrepreneurs with Silicon Valley, permitting a flow of capital, technology, marketing know-how and R&D into the Chinese companies. Through various science park-based 'clusters', the transnational networks intersect with the local and national networks to further support the diffusion of innovation. The Chinese success in repatriation contrasts starkly with the Indian case, where the return of professionals is only a 'trickle' (Saxenian 2001).

The government plans to boost software exports from $130 million in 1999 to $1 billion in 2004 by offering tax breaks and access to cheap capital and by relaxing rules on sending employees abroad (Ju 2001). China has so far concentrated on the domestic market, in contrast to India. Responding to the Chinese competition, many Indian companies have started to open up development centres on the Chinese mainland to re-route low-end activities like coding and maintenance. Some Indian companies are even hiring Chinese programmers who are less expensive than Indians (by approximately 15 per cent) and their language background makes them more suitable to support the efforts of Indian firms to penetrate Japanese projects. The Chinese workers, however, have lesser experience in the areas of systems integration and project management.

The Chinese market is divided into system software (12 per cent), application software (63 per cent) and supporting software (25 per cent). China has about 400,000 people employed in the software industry spread over the economically developed regions and coastal areas such as Beijing, Shanghai, Shenzhen, Dalian, Shenyang, Fujian and Zhuhai. Beijing is set to become China's largest software production centre with the municipal government approving 221 new software companies in 2000. The Zhonguancun science and technology park in Beijing represents China's Silicon Valley and is home to IBM and Microsoft. There is also the potential for Hong Kong to serve as a hub for outsourcing to the Chinese mainland as it is considered less risky and there are more English-speaking people. Already, the Chinese cities of Guangzhou and Shenzhen are host to a growing number of satellite offices for Hong Kong software companies. Some current barriers to growth include factors similar to those in Russia: poor English-language capabilities outside of Hong Kong, weak understanding of Western business culture and a poor reputation for IPP. To a greater extent than in India, Chinese companies have a vast domestic market to concentrate on and the Chinese official machinery is making efforts to address these limitations, making China a potentially significant future player in the GSW marketplace.

Figure 1.1 provides a brief summary of the major facets of these major software exporting countries, showing the spatial organization of the major centres.

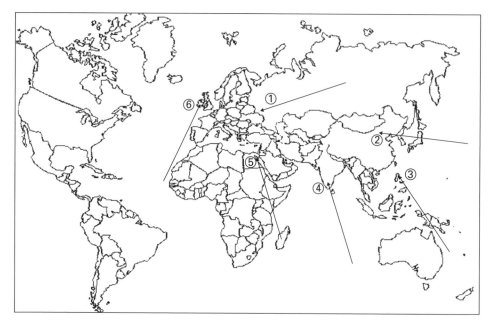

① **Russia**
- Centres at St Petersburg, Moscow and Novosibirsk
- US$ 60–100 million software exports (2000) 40–60 per cent yearly growth rate
- Full range of IT services, strength in high-level scientific work

② **China**
- Centres at Beijing, Shanghai, Shenzhen, Dalian, Shenyang, Fujian and Zhuhai
- Software industry has grown 20 per cent a year to a US$ 2.16 billion industry
- Predicted software exports from $130 million in 1999 to $1 billion in 2004
- Presently mainly coding contract activity

③ **The Philippines**
- Centres mainly around Manila
- Fast-growing industry estimated at US$ 200 million (2000)
- A range of IT-enabled services

④ **India**
- Centres at Delhi, Bangalore, Pune, Hyderabad and Chennai
- About $8 billion worth of software for foreign clients (2001) accounted for 10 per cent of India's total exports
- 40 per cent growth per annum, predicted $50 billion industry by 2008
- Full range of IT services and IT-enabled services

⑤ **Israel**
- Centres at Tel-Aviv, Haifa and Jerusalem
- Output of $1.3 billion in the first half of 2000, a 40 per cent jump from 1999
- Israel's specialization is in packages especially serving the Internet and the communications sectors

⑥ **Ireland**
- Total exports were valued at £4.2 billion (1998)
- Growth at 20 per cent per annum for most of the 1990s
- 80 per cent software exported
- Most activity around the major centre of Dublin
- Ireland is a centre for full range of IT services, package customization and IT-enabled services

Figure 1.1   Global offshore software production centres

1.5 Future challenges and opportunities

We have described how firms, supported by their respective countries and industry associations, are vying with each other to provide 'outsourcing' services globally. These outsourcing providers have to compete with other entities that provide 'insourcing' and 'nearsourcing' services. 'Insourcing' refers to in-house IT departments competing with external firms for providing IT services to their own organization (Hirschheim and Lacity, 2000). 'Nearsourcing' refers to firms in countries such as Mexico and Barbados who are competing primarily for the US markets based on the value proposition of greater geographical and cultural proximity to the USA as compared to outsourcing providers like India (Lapper and Tricks 1999). Nearsourcing countries also become cheaper alternatives for locating Indian programmers for projects in North America and for working around the problem of visa restrictions. (Mexico, for example, increased its offshore IT income by more than 80 per cent to $30 million in 2000.) Microsoft is one of a number of international companies outsourcing to a Mexican company, Softek, where engineers earn approximately $1,000 a month, less than a third of the pay in the USA but double that for the equivalent worker in India. Venezuela and Brazil are also potential nearsourcing options for US firms. Carmel and Agarwal (2002) also describes Canada as a potential nearsourcer as it has a 30 per cent cost advantage relative to the USA, on a par with Brazil and Ireland. Our definition of nearsourcing is not simply restricted to transatlantic relations as there is already evidence of Chinese companies serving Indian and Japanese firms. Both insourcing and nearsourcing will potentially be strong competitors to GSW in the future as firms learn about the limits to doing software work at a distance.

The various global options that firms adopt to meet software needs also introduce complexity when deciding on a specific country and firm. A manager has to process a multitude of information, commonly available through electronic sources whose credibility is often unverifiable. This decision is influenced by factors such as the type of work to be outsourced, the infrastructure in place and the perceived risks, strengths and weaknesses of the different vendors. To some extent, particular types of software development may be best suited to particular countries and firms, implying that companies need to develop a portfolio of outsourcing vendors from different countries. Microsoft has offshore arrangements in India, Russia and Israel. IBM's JavaBeans project involves centres in Riga, Minsk, Beijing, Bangalore and North Carolina. Nortel has its own development labs or alliances with companies in different countries including India, the UK, Canada, Israel and Vietnam. While this portfolio approach opens up a range of possibilities, it also creates extreme complexities, as managers have simultaneously to coordinate work in different cultures and work systems.

From the supplier side, the competition is intense. India faces stiff competition from Russia and China and to a lesser extent from Vietnam and the Philippines in the Far East.

The nature of competition is dynamic as companies are continuously seeking new markets and technologies. The more mature Indian software companies such as Infosys and Wipro, in their attempts to go up the value chain, are seeking to enter into consultancy-type contracts where they not only implement and solve problems, but also define them. Such mature companies do not see China, just entering the outsourcing arena, as a major threat but instead, view management consultancy firms like Pricewaterhouse-Coopers and Arthur Andersen as their key future competition. Initially, Chinese firms can find their competition coming from the new tier-two and tier-three Indian firms springing up in towns like Mysore and Mangalore which are satellite to Bangalore, and whose costs are lower than the 'top ten' such as Infosys and Wipro, and on a par with the Chinese firms. The future possibilities for growth will no doubt be shaped by the ability and capacity of firms to deal with the volatility and uncertainty of the GSW market. Countries such as India, which at the time of writing had a more than 60 per cent dependence on the USA, have the extremely complex task of reducing this dependence by penetrating new markets in Europe and East Asia. The Indian industry must also seriously reconsider their strategy for human resources (HR) development, as they cannot depend on people going continually to the USA. In 2001, a large number of expatriate programmers were 'benched' or returned to India, resulting in a slowdown in the recruitment of new graduates. While this setback is largely confined to the younger developers, there is the danger that in the long run this will make middle management hard to come by, leading to serious problems in developing project management capabilities.

As companies enter new domains and countries, they have the important task of developing strategies that blend adopting a global outlook and building on local strengths. A key challenge is to develop capabilities to handle complex communication and coordination problems in conditions of time, space and cultural separation. These problems are magnified in the absence of reliable infrastructure and in domains where user specifications are constantly subject to change. Mature project management abilities, especially the capability to deal with quality and time standards, will be another important determinant of where work is sourced. While outsourcing firms are becoming increasingly sophisticated in developing systems to create transparency in their operations, the demands and expectations of clients are constantly increasing. This implies a process of continuous learning, reflexivity and renegotiating the terms of the relationship. IP regulations, which may not have been a significant issue when doing lower-end and maintenance work, become crucial as suppliers try to go up the value chain and seek profit sharing and royalty-based arrangements. How well countries can revamp their legal systems and create confidence and trust among the investing companies will be an important factor in the future. There is a need to have realistic expectations of what can and what cannot be done in a distributed setting and to realize that ICTs on their own cannot 'make geography history'.

In summary, we have discussed the nature of organizational forms, the kind of work and the global trends that surround GSW. These key features, which are summarized in

Table 1.1 Key features defining and shaping GSAs

Defining themes	Key features
Nature of organizational relationship: GSA	Network structure spread across temporal, spatial, organizational and national boundaries
	A new breed of smaller 'born global' firms
	Central role of ICTs in coordinating network activities
	ICTs necessary but not sufficient condition for effective coordination
Nature of work done: GSW	GSW reflects features of both services and manufacturing
	GSW involves varying level of intellectual and creative input
	There are varying levels of scalability
	Magnified influence of social and human issues
	Rapid technological and organizational change in GSW requires continuous reflexive monitoring
Nature of global trends	Multiple options for customers of GSW in terms of countries and firms
	Heightened competition for suppliers in a rapidly expanding marketplace
	Competition from 'nearsourcing' and 'insourcing' vendors
	Need to develop blend of global and local approaches to management
	As GSW becomes increasingly common, there is the need to understand the process of how a GSA relationship works in practice

table 1.1 are interconnected and together shape the future challenges and opportunities in the management of GSA arrangements.

While the practice of GSW has come a long way in developing strategies and approaches to deal with the complexities of managing GSAs, the field of information systems research lags far behind in understanding how these relationships work in practice. Prior research in this area has been limited more to national than to global settings, with a focus primarily on services such as the maintenance of help desks or call centres and transaction systems such as airline ticketing. Another feature of this research has been its focus on understanding how various factors influence managers' decisions to enter outsourcing contracts. Some of these factors include communication and coordination problems, local context and cultural issues and unrealistic expectation of managers (Apte 1990; Vitalari and Wetherbe 1996; Carmel 1999; Gupta and Raval 1999). While the knowledge of these factors sensitizes us to important issues in deciding the feasibility of entering into outsourcing arrangements, they are rather static in nature and have limited capability to explain how a GSA relationship that is inherently processual can unfold over time.

This book is written with the assumption that GSW has to a great extent been normalized in business practice in that it is seen to be advantageous to both vendors and customers. The relevant question then is not 'whether we should enter into GSA relationships but how can we manage them more effectively?': the challenge is to focus on how these relationships play out in practice over time and what we can do to make them more effective. With these challenges in mind, the guiding question in this book is how GSA relationships evolve. This question is addressed through empirical investigations of seven longitudinal case studies whose analyses are informed through a variety of theoretical perspectives. A unifying backdrop to these analyses is the context of globalization within which these relationships are situated. As a starting point, therefore, in chapter 2 we first discuss the analytical relationship between the processes of globalization and then identify its defining features. These features provide the content of investigation of the subsequent case studies.

BIBLIOGRAPHY

Note: This introductory chapter also includes some further and background reading on the whole global IT outsourcing field.

The publisher has used its best endeavours to ensure that the URLs for external websites referred to in this book are correct and active at the time of going to press. However, the publisher has no responsibility for the websites and can make no guarantee that a site will remain live or that the content is or will remain appropriate.

ACCR (2001). *White Paper on Offshore Software Development in Russia*, American Chamber of Commerce in Russia, 21 March, http://RussianSoftwareDev.eSolutions.ru

Apte, U. (1990). Global outsourcing of information systems and processing services, *The Information Society*, 7, 287–303

Ariav, G. and Goodman, S. E. (1994). Israel: of swords and software plowshares, *Communications of the ACM*, 37, 6, 17–21

Bartlett, S. C. and Ghoshal, S. (2000). *Transnational Management: Text, Cases and Readings in Cross-Border Management*, Boston: Irwin McGraw-Hill

Bjerknes, G., Bratteteig, T. and Espeseth, T. (1991). Evolution of finished computer systems: the dilemma of enhancement, *Scandinavian Journal of Information Systems*, 3, 25–46

Carmel, E. (1999). *Global Software Teams: Collaborating Across Borders and Time Zones*, Upper Saddle River, NJ: Prentice-Hall

Carmel, E. and Agarwal, R. (2002). The maturation of offshore sourcing of IT, MISQ executive, 2, 65–78.

Castells, M. (1996). *The Rise of the Network Society*, Oxford: Blackwell

Cochran, R. (2001). Ireland: a success story, *IEEE Software*, 18, 2, 87–95

De Fontenay, C. and Carmel, E. (forthcoming). Israel's Silicon wadi: the forces behind cluster formation, in T. Bresnahan, A. Gambardella and A. Saxenian (eds.), *Building High Techclusters*, Cambridge: Cambridge University Press. http://www.american.edu/ksb/mogit/carmel.html

Friedland, R. and Boden, D. (eds.) (1997). *NowHere Space, Time and Modernity*, Berkeley: University of California Press

Friedman, A. (1989). *Computer Systems Development*, Chichester: Wiley

Gupta, U. and Raval, V. (1999). Critical success factors for anchoring offshore projects, *Information Strategy*, 15, 2, 21–7

Gurbaxani, V. (1996). The new world of information technology outsourcing, *Communications of the ACM*, 39, 7, 45–6

Hamlin, M. (2001). Measuring the new economy, *South China Morning Post*, 5 February, http://www.teamasia.com.ph/Media2001/02052001A.htm

Heeks, R. (1995). Global software outsourcing to India by multinational corporations, in P. Palvia *et al.* (eds.), *Global Information Technology and Systems Management: Key Issues and Trends*, Nashua, NH: Ivy League Publishing

  (1999). Software strategies in developing countries, *Communications of the ACM*, 42, 6, 15–20

Hirschheim, R. and Lacity, M. (2000). The myths and realities of information technology outsourcing, *Communications of the ACM*, 43, 2, 99–107

IPR (2000). Strategic Business Database, *Globe*, 24 August (Israeli newspaper)

Ju, D. (2001). China's budding software industry, *IEEE Software*, 18, 3, 92–5

Kobitzch, W., Rombach, D. and Feldmann, R. (2001). Outsourcing in India, *IEEE Software*, 18, 2, 78–89

Krill, P. (2001). United States increases offshore outsourcing, *InfoWorld*, 23, 10, 16

Lakaeva, I. (2000). Computer software market for offshore software development, *US Commercial Service* (Moscow), http://www.bisnis.doc.gov/bisnis country/000829RusSoftDev.htm

Lapper, R. and Tricks, H. (1999). Nearshore contracts flow Mexico's way, *Financial Times*, 17 May, 16

McFarlan, F. (1995). Issues in global outsourcing, in P. Palvia *et al.* (eds.), *Global Information Technology and Systems Management: Key Issues and Trends*, Nashua, NH: Ivy League Publishing

Moore, S. (2001). Offshore and nearshore outsourcing options, *Giga Information Group*, 29 August, http://www.gigaweb.com

NASSCOM (1998). Software industry in India, National Association of Software and Service Companies, New Delhi, India

  Software Industry in India, National Association of Software and Service Companies, New Delhi, India

O'Riain, S. (1997). The birth of a Celtic tiger, *Communications of the ACM*, 40, 3, 11–16

Ritzer, G. (1996). *The McDonaldization of Society*, 2nd edn., Thousand Oaks, CA: Pine Forge Press

Saxenian, A. (2001). The Silicon Valley connection: transnational networks and regional development in India, Taiwan and China, Final Report for the project *The Context of Innovation of the Information Technology Industry*, University of Pennsylvania Institute for the Advanced Study of India, New Delhi, India, August

Teubal, M. (2001). The successful development of small high-tech firms in the Indian economy, Final Report for the Project *The Context of Innovation of the Information Technology Industry*, University of Pennsylvania Institute for the Advanced Study of India, New Delhi, August

Vitalari, N. and Wetherbe, J. (1996). Emerging best practices in global systems development, in P. Palvia *et al.* (eds.), *Global Information Technology and Systems Management: Key Issues and Trends*, Nashua, NH: Ivy League Publishing

Willcocks, L., Fitzgerald, G. and Lacity, M. (1996). To outsource IT or not? Recent research on economies and evaluation practice, *European Journal of Information Systems*, 5, 143–60

Willcocks, L. and Sauer, C. (2000). High risks and hidden costs in IT outsourcing, *Financial Times*, Mastering Risk Supplement, 23 May

FURTHER AND BACKGROUND READING

Apte, U. and Mason, R. (1995). Global disaggregation of information intensive services, *Management Science*, 41, 7, 1250–62

Arora, A., Arunachalam, V. and Asundi, J. (2000). The globalisation of software: the case of the Indian software industry, http://www.heinz.cmu.edu/project/india/index.html

Carmel, E. and Agarwal, R. (2001). Tactical approaches for alleviating distance in global software development, *IEEE Software Special Issue on Global Software Development*, 18, 2, 22–9

Clark, T., Zmud, R. and McCray, G. (1995). Transforming the nature of business in the information industry, *Journal of Information Technology*, 10, 221–37

Coe, N. (1999). Emulating the Celtic tiger? A comparison of the software industries of Ireland and Singapore, *Journal of Tropical Geography*, 20, 1, 36–55

Heeks, R., Krishna, S., Nicholson, B. and Sahay, S. (2001). Synching or sinking: trajectories and strategies in global software outsourcing relationships, *IEEE Software Special Issue on Global Software Development*, 18, 2, 54–61

Heeks, R. and Nicholson, B. (2003). Software export success factors and strategies in developing transitional economies, IDPM, University of Manchester, Working Paper, 12, http://idpm.man.ac.uk/wp/dc

Kitov, V. (2001). Software firms aim to copy India's success, *The Russia Journal*, 8–14 June, 12

Philippines data: http://www.e-servicesphils.com

Rajkumar, T. and Mani, R. (2001). Offshore software development: the view from Indian suppliers, *Information Systems Management*, Spring, 63–72

Willcocks, L. and Lacity, M. (2001). *Global Information Technology Outsourcing*, Chichester: Wiley

2 Globalization and global software work

2.1 Introduction

Global software alliances (GSAs) are work configurations in present-day processes of globalization. GSAs involve the restructuring of work arrangements across time and space based on the underlying assumption that the experience of the 'here' and 'now' of software development can be largely superseded by action at a distance. GSAs represent the state of the art in software development because of the inherent potential they provide organizations for standardizing, measuring and coordinating distributed resources, including HR and resources of time and space. GSAs reflect Mowshowitz's (1994) vision of a future paradigm of management based on the principle that the 'best' and 'cheapest' HR can be 'switched' as required and assembled from around the globe and coordinated through the use of ICTs and standardized methodologies and processes. These underlying principles provide the basis for popular claims that software development can 'follow the sun' with round-the-clock operation by exploiting time-zone differences allowing cycle-time reductions in handling work between sites located in different parts of the world. Such a vision of GSAs as an arena in which the 'switching of resources' takes place is no longer merely a speculative phenomenon of the future, but one that is being actively attempted by firms globally, with varying degrees of success.

The aim of this chapter is to understand the mutual interconnection between processes of globalization and global software work (GSW). We start by outlining a conceptual stance on the nature of this interconnection, which then permits us in the next section to examine its implications for the structure, processes and nature of risks that underlie GSW. Against the background of this globalization–GSA relationship, we set out to introduce the case studies discussed later in the book that provide further insight into the nature of the relationship.

2.2 GSAs as 'models of' and 'models for' globalization

Drawing upon Geertz's (1973) terminology, we conceptualize GSAs as serving as a metaphor for both 'models of' and 'models for' globalization, reflecting and acting as agents of globalization processes. Such a conceptualization helps to reflect the complexity of this mutual relationship that requires both the immediacy of face-to-face co-presence and action at a distance. There is an inherent and ongoing tension as global configurations oppose and also support various local and regional groupings. The interplay between proximity and distance is paradoxical, requiring open and trusting relations that are normally seen as a function of proximity and not distance.

Processes of globalization are interconnected to GSA processes at various levels. At the industry level, there is the integration of the communications and computing sectors or redefinitions in software development approaches created through the Internet and related technologies. At the more macro level, these global changes have immediate and intensified implications for the technology focus of the GSA, the organizational forms and the markets addressed. For example, one Indian firm doing well in the Electronic Design Automation (EDA) market decided to shut down that area of work to focus on telecommunications, which they saw to be the future area of growth. This decision, influenced by a number of global reasons, had immediate implications for the structure and thrust of their GSA arrangements. It affected the individual level in terms of the kind of people hired and the type of technological skills required; it also affected which geographical regions to focus on, further implying a need to develop new language and cultural understanding. Another example is the slowdown of the US market in 2001, forcing firms throughout the world to seek alternative markets in Europe and East Asia so as not to perish by putting 'all their eggs in one US basket'.

While the interconnection between the global and individual level may not be something unique to today, what is different is the *speed and reversibility of effects*. The changeover from EDA to telecommunications mentioned above was extremely swift and decisive. In between two phases of interviews in the same year, we found a firm to have made this transformation, including its name, Web sites, mission statement, organizational structure and espoused values and assumptions. The reversibility of effects comes from the knowledge-intensive nature of work and the significance of the individual developer. The Indian firm above could not have made this transition if it were not confident of recruiting, retraining and retaining professionals with telecommunications expertise. As compared to accounting and office management, it is much harder to build control systems in GSAs that transcend the individual, especially in highly knowledge-intensive and relatively new areas such as mobile telephony. Managing 'key HR' is an important task in GSAs, and the reversibility of effects on how global work is structured arises with individuals deciding to apply their knowledge capital in alternative settings.

Writers like Giddens (1990) and Castells (1996) describe a distinctive feature of contemporary life as the increasing interconnection between the two extremes of *globalizing influences* and *personal dispositions*. Identity, both individual and organizational, becomes a key element shaping and defining the nature of these interconnections. The dynamics of globalization imply new influences on identity that get hybridized in different ways. For example, a typical Indian software firm will undertake projects in

North America, Europe, and East Asia, implying a need for developers to slip in and out of different technical, social and cultural experiences. These constant movements force individuals reflexively to monitor their preparedness and take corrective action. Radical and constant changes in cultural and work contexts are associated with trans-formations in self-identities, and are manifested in many forms, including the constant movement of software developers or the adoption of American life style in Bangalore and Hyderabad.

Intensification of the interconnections of processes over time and space, reflexiv-ity of knowledge and transformations of individual and organizational identities are all characteristic of ongoing processes of globalization. GSAs reflect these character-istics and also shape them in particular ways. Indian software companies, based on the learning and experience gained through GSAs with North American companies, are now actively exploring Japanese and other East Asian markets, adopting their own distinctive form of hybridized ways of software management. These hybridized management styles will no doubt be further redefined, as the 'Japanese flavour' of working is incorporated and creates other situated meanings of globalization. We thus argue against the 'globalization as homogenization' thesis (for example, Hannerz 1991) and take the perspective of authors such as Robertson (1992) who use the term 'glocalization' to describe globalization as the 'interpenetration of the universalization of particularism and the particularization of universalism' (1992: 100). Writing in the context of IT and the world of work, Walsham (2001) similarly argues for globalization and diversity:

> However, the change processes are not seen as uniform in their effects, and organizations and societies are likely to remain distinct and differentiated, although increasingly interconnected through a process that has been labelled as glocalization. This arises through indigenization, meaning the selective appropriation of new ideas by different individuals, organizations and societies. (2001: 32)

Conceptualizing GSA relationships in the context of globalization sensitizes us to some of these meta-level dynamics and tensions that characterize such global phenomena. However, such theories of globalization, for example those of Giddens and Castells, while helping us to 'view the landscape below from . . . a [remote] plane' (Walsham 2001), are limited in explaining the micro-level dynamics of particular organizational processes. 'High-level' insights generated from social theories need to be integrated with in-depth, micro-level and processual analysis of particular work arrangements. To understand this integration process, we now outline the broad trends in some current sociological theories of globalization and subsequently relate them to the analysis of GSAs.

2.3 Theories of globalization

A sociological perspective on globalization, given the primary focus on 'understanding the process by which a GSA relationship evolves over time', broadly informs our analysis of GSAs. This evolution is primarily a social process involving a relationship between people, organizations and technology within a global setting. We draw upon the writings of three contemporary and influential social theorists – Anthony Giddens, Manuel Castells and Ulrich Beck – to situate GSAs conceptually within globalization, especially to analyse the *structure, process* and *nature of risks* of GSAs.

Structure of GSAs

Castells (1996) uses the 'network society' as a metaphor to describe contemporary social structure as made up of networks – and more specifically, *informational networks*. There are two key elements in Castells' analysis – networking and technology. Networks are made up of interconnected nodes with no centre, exemplified in global financial networks, production and consumption organized around the network enterprise, the global criminal economy and various other networks that exert joint influence on global flows of power and wealth. Castells emphasizes the importance of such networks:

Because networks are extremely efficient organizations, they eliminate, through competition, alternative structures, so their logic expands. Because they operate in a globally interconnected environment, they diffuse unevenly, throughout the planet, blurring institutional and cultural boundaries, and focussing exclusively on their instrumental performance. Networks are the carriers of globalization. (Castells 2001: 5)

In chapter 1, we discussed how the structure of GSAs can be conceptualized using this 'network' metaphor that includes a set of software development nodes (of users and providers in the USA, India, Ireland, Israel, etc.) linked together through the extensive use of ICTs and various standardized processes. In this network structure, an informational logic is fundamental and characterized by two ongoing processes. First, firms try to situate themselves within a network that can help increase the informational content of their activities – for example, software firms increasingly rely on the Web for meeting their recruitment needs and so build networks with various employment agencies that maintain relevant personnel databases. Secondly, firms try to network with organizations specializing in R&D, engaged in generating new knowledge, and integrating this knowledge into internal production processes to enhance the value of the firm. To permit this external and internal networking, firms need to have a structure that is flexible and supportive of ongoing learning processes. Developing such networks is not without inherent contradictions: while cost considerations drive software production to areas where labour costs are lowest, software development is seen by many to be more effective in conditions of co-location of developers in proximity to the user.

While networks are not a new phenomenon, what distinguishes GSA networks from what existed in the past is the extent of use of ICTs, which helps to impart flexibility to the software development process in ways not possible before. While past networks could perform tasks of limited functionality, Castells (2001) argues that present-day networks (like GSAs) permit more complex tasks to be undertaken:

> They [earlier networked enterprises] failed in their coordination functions; they could not master large resources, and marshal them in one particular direction. Centred, hierarchical organizations were much more efficient at mobilizing resources and executing the order. New information technologies changed all this. Suddenly networks could at the same time coordinate decision-making and decentralize execution of shared tasks. They could issue orders, and receive feed back in real time. They could be multi-directional and interactive. (2001: 5)

The aspects of interactivity, adaptability, coordination and decentralization of execution tasks, which Castells describes above, are fundamental to GSAs. Without ICTs, distributed development would be impossible, since programmer teams would not be able to coordinate almost instantaneously with each other to clarify requirements, deal with bugs, monitor quality levels and build trust in the skills of other developers without being able to see them face to face. There is a technological determinism in the argument that ICTs fundamentally provide the capability to organizations to coordinate and decentralize the execution of tasks and GSAs would never have been possible if today's technological infrastructure (such as the Internet and the Web) did not exist. However, it is not totally deterministic since the evolution of GSA networks takes place within particular historical and social contexts that shape the dynamics of their evolution in particular ways. Large MNCs typically have the power to make independent investments to establish infrastructure, and lobby for the implementation of standards that are advantageous to them. This power is evident in that the first entrants into the GSW arena were large corporations like Texas Instruments, Nortel and Oracle. But as the networks diffused and various cost sharing models evolved, we find the active presence of smaller organizations, including those based in developing countries like India. These firms challenge the traditional power dynamics by virtue of their capacity to leverage intellectual capital using a networking logic that 'induces a social determination of a higher level than that of the specific social interests expressed through the networks: the power of flows takes precedence over the flows of power' (Castells 1996: 469). The network structure helps to shift power dynamics and redefine patterns of communication and cooperation among institutions.

GSA processes

Anthony Giddens' (1999) 'runaway world' serves as a useful metaphor to emphasize the speed and 'out-of-control' nature of GSA processes. Processes of change and the interdependency of systems influence the status of knowledge, contributing to the sense of a juggernaut that is out of control. Giddens argues that this intensification needs to be understood at an institutional level and in its interconnection with the individual level. ICTs facilitate this intensification, and serve as 'disembedding mechanisms', enabling formalization, codification and redistribution across other time and space domains. In GSAs, the use of software development methodologies reflects these disembedding mechanisms that are transported across time and space. However, there are often problems in applying these methodologies in practice because of the complexity inherent in codifying different forms of knowledge.

In conditions of intensified global interdependencies and reflexivity of knowledge, no firm can take it for granted that any established GSA will go on forever. Each actor is constantly acquiring new knowledge about technologies, markets and organizational arrangements. Each actor is also coping with global events (like the 2001 US slowdown) that places new demands on the relationship. Changes in globalization are not external and 'out-there' but something internal and defined by internal actions such as a partner deciding to terminate the relationship in order to exploit new market opportunities. Such institutional changes resulting from new markets or technologies have implications at the individual level, as evidenced in the extensive and continuous global movement of software developers. Attracting and retaining human talent is one of the key management challenges of GSAs. Giddens (1994) describes the current processes of capitalism to be characterized by competitive markets, the commodification of labour power and the insulation of the economic system from the political arena. An example of this insulation is reflected in Germany's decision in 2001 to allow the entry of foreign programmers under the 'green-card' scheme. This scheme was implemented for economic and resource considerations despite popular political protests that Germans would lose employment possibilities.

Another key aspect of current processes of globalization concerns the 'reverse effects' that reflect the mutual impact of local and global events. No longer are local events shaped only by global happenings, but the global is also shaped by local happenings. The 'green-card' example referred to above is a case in point. Trying to attract foreign software professionals opened a wider debate about Germany's immigration policies in general. At another level, software development activities taking place in an organization in India are shaped by methodologies and processes developed in the USA – for example, systems for quality control. However, through the incorporation of these standardized processes into their work practices, Indian firms become more 'global' and this helps to encourage other foreign firms to move their software development activities to India. Large-scale relocation of work of this kind can potentially lead to unemployment or underemployment conditions in the country from where work is relocated. Events occurring locally are thus both shaped by and also shape global happenings.

Risks in GSAs

The structure and processes of GSAs both emphasize the unique nature of risks in such arrangements. These risks can be examined through Ulrich Beck's (1992) metaphor of the 'risk society' that emphasizes the changing nature of relations between social structures and human actors in contemporary life. Beck draws our attention to the risks inherent in social structures today, arising in particular from their global nature, their interdependence and the political and social nature of the process by which risk is defined and globally diffused. GSAs, conceptualized as work configurations in a risk society, draw attention to the inherent risks and how they are redistributed globally across time and space.

Risks come in many forms and affect multiple levels including nations, industries, organizations and particular individuals. An example of the nature and interdependence of these risks is seen in the manner in which the 2001 US slowdown affected the entire GSW marketplace, including cancellation of projects, laying off of personnel and firms seeking new markets in Europe and East Asia. This slowdown was magnified by the 11 September attacks in New York that introduced new risks in travel and international business. The interdependence of risks comes in many forms, including political, economic as well as geographical, where events in Manhattan have profound global implications. Another important feature is the manner in which risks are politically and socially constructed. The continuous CNN broadcasts on the 'war against terror' influence our construction of what terrorism is and the associated risks.

GSAs, by virtue of their globally diffused network structure, reflect and transmit contemporary risks. Managers need to respond at short notice to these risks and redefine structure, address new markets, learn about new technologies and build buffers to withstand shocks arising from cancellation of projects. Risks are not always external but can also be internal, as GSAs depend on a large and complex technological infrastructure for their survival. Breakdowns in infrastructure can disrupt the project where even small downtimes can be fatal. As Beck argues, a fundamental concern of organizations in the risk society is not the redistribution of wealth, but the redistribution of *risks*, and seeking new geographical markets and technologies are strategies being used to redistribute risks. The risks of success or failure of projects are inherently incalculable, particularly the risk of whether a piece of software that has been built partly in India and Russia will function in a bug-free manner. Despite the various quality certification processes – for example, ISO 9000 and CMM Levels 1–5 – such risks of software malfunctioning remain inherently incalculable. The unpredictable nature of the risks heightens the potential for unintended consequences.

Risks also come with the new kinds of choices that GSW permits; for example programmers can choose where and with whom to work. Beck describes the current nature of 'individualization' as an 'I am I' structure that implies liberation at one level, and at another a loss of stability and 're-embedding' into other contexts. The very medium that enables individualization also brings about standardization because individual situations are thoroughly dependent on the global labour market that operates through standardized capitalist templates. Money both individualizes and standardizes. Beck thus argues that we are confronted with a situation of 'institutionally dependent individual situations' that '[deliver] people over to an external control and [develop] standardization that was unknown in feudal times'. While on one hand developers have the possibility to select employment options from a set of global alternatives, at another

Table 2.1 The 'model of' and 'model for' relationship

Macro-level aspects	Features	Relation to micro-level themes
Structure	Network configuration Informational flows Power of flows Decentralization of execution Role of technological innovation	Tensions of space and place Issues of power and control
Process	Intensification Unintended effects Reverse effects	Transformations of identity Tensions of standardization
Nature	Transferring risks Risks and software development Risks inherent in new knowledge systems Destandardization	Complexity of knowledge transfer Language and culture

level these choices are made within standardized templates of salaries, stock options, protection against termination and other benefits. However, these choices are not insulated from the risks that institutions have to face, such as the situation resulting from the meltdown of the dotcom phenomenon in North America in 2000. The theoretically attractive choice of going to North America comes with risks that are fundamental to the capitalist structure within which these choices are offered.

The metaphor of the risk society helps to conceptualize the nature of risks that GSAs have to contend with arising from their structure, processes and the global context within which they are situated. While GSAs have constantly to seek to develop strategies to redistribute the risks they deal with, they are also *carriers of these risks* into the various domains where they operate. Global risks are just the part of everyday existence of GSAs that cannot be avoided and must be confronted on a continuous basis.

In table 2.1, we summarize key features of the structure, process and nature of GSA risks that reflect the different aspect of the 'model of' and 'model for' relationship between GSAs and globalization. We relate these features to various 'micro-level' themes that are drawn from the literature on globalization and are also themes that we have identified through our empirical analysis to be of prime importance in understanding the process of evolution of GSAs. We elaborate on these themes in the following section.

2.4 Micro-level themes in GSW

We discuss six themes identified as a result of both a 'top-down' analysis of literature and a 'bottom-up' investigation of empirical material. 'Top-down' in that these themes find extensive mention in the various writings on globalization, some of which were discussed earlier. 'Bottom-up' because these themes also reflect patterns observed in the empirical analysis of different case studies. Arising from this 'top-down' and 'bottom-up' interplay, which van Maanen (1989) describes as a conversation between conceptual understanding and empirical analysis, we introduce these themes and draw upon them in subsequent chapters to analyse how GSA relationships evolve over time.

Tensions of place and space

'Space' and 'place' have long been concepts of analytical interest to human geographers, and in recent times find central mention in current writings on globalization. Giddens (1990) analyses contemporary life through changes in the conditions of social interactions, from face-to-face or 'place-based' settings in traditional societies to that of absence or 'space'-based settings in current times. Emphasizing this surge of interest in space and place, Dirlik (1998) writes:

> The past decade has fortunately seen the eruption of place consciousness into social and political analysis. Place consciousness is closely linked to, and appears as the radical other of, globalism. (1998: 8)

'Space' and 'place' serve as metaphors to understand the relation between material practices and the physical and electronic domains within which they occur. While 'place' is associated with a person's sense of boundedness and particularity, 'space', in contrast, refers to a sense of universal and the abstract. A place is a space to which meaning has been ascribed and represents a psychologically meaningful material space. Instead of thinking of places as physical areas with boundaries, they can be conceptualized as *articulated movements in networks of social relations and understandings*. GSW, inspired by the discourse of globalization, assumes a production, capitalist oriented logic that software development can occur in any location (space) as long as labour is available and development costs are low. In the practice of GSW, this assumption stands in tension with organizational and individual needs for place-based work that emphasizes the local and particular. While GSW seeks to operate with a placeless logic, its everyday work practices are primarily place-dependent. Even though firms assume that they can hire people from any part of the world, to actually be able to do so they need to develop an understanding of social life in which place is crucial. To retain an Indian programmer who typically prefers to go to the USA, the firm will need to understand his/her family and social structure and what it means in that context for him/her to go to the USA. Such an understanding is fundamentally place-based, since an individual is not only a member of an organization but also belongs to a family, a community, a broader society and nation. These different memberships introduce their own norms and values that are fundamentally place-based and difficult to transcend.

Space-based work is an underlying assumption of GSW, which is in constant tension with the everyday work practices of the relationship at both the individual and organizational levels that vary with the different stages of evolution of the relationship.

We use the lens of this space and place tension as a basis to analyse the case of GlobTel–MCI discussed in chapter 6.

Complexities of knowledge sharing

Software development is a knowledge-intensive activity, which is further intensified in the GSW context because, in addition to the product domain knowledge required, further knowledge needs to be developed about various contextual issues such as language, national policies, etc. that shape the process of GSW. A key feature of globalization concerns 'disembedding mechanisms' (Giddens 1990) which refer to the processes by which local practices are codified, lifted out of a particular context and re-articulated in other global domains. To permit this 'lifting-out' process, local interactions and knowledge need to be formalized, codified, disseminated and re-embedded in other domains. This concept has interesting implications for understanding the issue of 'knowledge transfer' and its 'sharing' in GSW, where typically large and distributed organizations have to confront local variations and definitions of knowledge in their alliances in different countries. Complexities arise at various stages of the GSA process, including how knowledge is understood, codified, disembedded, transferred across time and space and re-embedded in other contexts.

Distributed software development fundamentally requires alliance partners to develop a shared understanding of each other's products, processes and work practices. For this, the outsourcing firm requires the transfer of various forms of knowledge to its partners to enable software development to take place. There are various factors that make this process extremely problematic, arising in particular from issues of tacit forms of knowledge and the background knowledge that is required to develop such understanding. Much of the current literature on knowledge management and transfer adopts a rather functionalist position, assuming knowledge to be objective, tangible and therefore transferable between humans and organizations through the use of ICTs. This literature has been a subject of recent critique and over the years information systems researchers have established the problematic nature of transferring knowledge between users and developers even in conditions of co-location. A lot of the knowledge held by the software developers is tacit, and transferring that in conditions of time, space and cultural separation is not problem-free.

To address issues of the subjective nature of knowledge and the social and individual nature of the processes by which knowledge is articulated, transmitted and understood, we adopt an interpretivist position that emphasizes the epistemology of practice as opposed to that of possession. The concept of *community of practice* (Lave and Wenger 1993) has been established as an effective way to address some of these issues in face-to-face settings. This concept is further developed with the work of Brown and Duguid (2000), who emphasize the process through which groups of practitioners are able to muster collective 'know how' and share this with colleagues in the course of practice. For Brown and Duguid, knowing is a kind of knowledge inseparable from action and practice is action that draws upon meanings created within a specific social context. They distinguish between *knowledge* and *knowing*, but these two concepts are in a 'generative dance' where knowledge disciplines knowing and knowing produces and reproduces knowledge. Knowledge is thus situated as both a practice and an object (Schultze 1998: 163).

GSW faces the extremely complex challenge of trying to develop 'universal knowledge' systems while dealing with the variations in the local, including firms and individuals. Our approach is to extend the community of practice idea to GSA to form an understanding of the relationship between GSW and knowledge, and the complexities that arise in sharing 'know how' with colleagues located at remote locations who have situated individual or collective background understanding. The effectiveness or otherwise of these processes of knowledge transfer significantly shapes the evolution of the GSA relationship over time. We consider the different kinds of complexities associated with knowledge at various points in the relationship, from initiation where the alliance is formed in an offshore location to the growth stage where trust is augmented by successful knowledge sharing. Effective knowledge sharing is a key to growth and maturity in GSAs, and can help to move the relation up the value chain to closer collaboration. Failure to do so can similarly have the opposite effect. The dynamics of knowledge transfer and its relation to the process of growth of a GSA is discussed in chapter 7 using the case of Sierra, a UK firm, and its relationship with its Indian subsidiary.

Tensions of standardization

As a strategy of globalization, many firms have established offshore development centres in different parts of the world to enable distributed software development. Bartlett and Ghoshal (2000) describe this as a 'global strategy' where worldwide activities are closely coordinated through central control from headquarters, gaining benefit from a standard product design and global-scale manufacturing. A dominant management perspective often taken is that the world could, and should, be treated as a single integrated market marked by similarities more than by differences. In GSW, standardization is a key feature in building and sustaining a relationship by homogenizing operations to the extent that the outsourcing and outsourced firms cannot be distinguished from each other. The scope of this standardization effort encompasses physical, technical and management domains. While the physical domain could concern the layout of offices, the technical domain involves using the same software development methodologies and the management domain refers to efforts to create a universal template that can be used to define, guide and evaluate management practices across the GSA.

To deal with complexities of knowledge, firms try to simplify tasks by standardizing various processes of knowledge transfer, such as how project reports are written and the criteria to judge the quality of a developer's work. These standardized systems, often codified in manuals and databases, serve as points of reference to coordinate activities of partners across time and space. Such attempts to standardize are rarely unproblematic, and are in tension with the need for flexibility at the 'local' level. While some degree of standardization is essential to enable global coordination, there is always the question of how much and what to standardize. While it may be relatively simple to standardize technical quality control methods, it is much harder to homogenize how managers from different backgrounds communicate.

Standardization takes place at multiple levels, including the choices of individuals to move from one organization to another. As management makes efforts to retain key individuals who are a globally sought-after 'commodity', they resort to offering various financial compensations and treating them as a 'key resource'. Individualization of these key resources is achieved through standardized market mechanisms of financial packages, career advancement schemes and stock options. The individual developer's aspirations and expectations find solace in compensation systems that are rooted in standardized and global market environments. Developers are individuals to the extent that they are primarily concerned with the kind of technologies they work with and how that adds to their career profile in the global marketplace. The individualization at this level takes place with reference to the standardized skill sets required in the global market.

In an environment as dynamic, interconnected and yet diffused, as that in which GSAs operate, trying to standardize is like shooting at a moving target. Constant movement takes place because of the dynamism of the global setting and the diversity of the partners involved. Hanseth and Braa (2001) argue that the hope of creating a universal standard is an illusion, akin to 'hunting for the treasure at the end of the rainbow'. Each time a standard is believed to be complete and coherent, it is usually discovered during implementation that there are elements lacking or incompletely specified while others have to be changed to make the standards work, which makes implementation difficult and incompatible – like arbitrary non-standard solutions. Efforts at standardizing are thus constantly being attempted within a 'non-standardizable' context, especially when dealing with management practices that are shaped by the response of individuals. However, some degree of standardization is essential, raising the complex challenge of how to forge a pragmatic balance between developing universal solutions and accounting for local reality.

The tension of standardization is ongoing over the course of the relationship: each stage places varying demands on standardization, in both the domain and extent of application. This dynamic interplay between standardization processes and the process of GSA growth is the lens through which to view the analysis of the case of the GlobTel–Witech relationship discussed in chapter 4.

Transformations in identity

Transformations of the self and their increasing interconnection with the institutional level are a dominant theme in current writings on globalization. Globalization needs to be interpreted in the context of the interconnections between groups and individuals with institutions. Giddens (1990), for example, discusses how interconnections between the vulnerable nature of knowledge and the placeless logic of modern life contribute to a feeling of 'existential anxiety' or 'personal meaninglessness' at the individual level. A response to these individual-level feelings of anxiety and insecurity comes through various expressions of identity. As organizations increasingly attempt to operate with a placeless logic, implying context independence, individuals primarily remain historically and biographically place-dependent. Castells describes this as dialectical relationship between the 'net and the self', which makes individuals experience a 'structural schizophrenia'. Castells (2001) argues that the power of identity is a defining feature in contemporary life:

It is true that new institutions could bridge the split between the net and the self, thus avoiding the potential disintegration of society. Indeed, this is my personal hope. In my analytical framework, this process is conceptualized as 'project identity', that is when social actors, on the basis of whichever cultural materials are available to them, build a new identity that redefines their position in society, and by doing so, seek the transformation of the overall social structure. (2001: 15)

The interconnection between individual identity and social structure raises the empirical question of how identity in GSAs at organizational and individual levels mutually relate. The dynamics of globalization means that identity at different levels is subject to new influences and is hybridized in different ways. Software companies involved in GSAs are being established in India by Indians educated in the USA, modelled on the Silicon Valley approach to innovation and creativity. The American identity is hybridized with Indian values such as 'eternal relationships' as a company strives to construct an organizational identity to face global competition while attempting to overcome local handicaps and build on local strengths. As these companies expand their operations into other markets, for example into Scandinavia, identity tends to become hybrid. Changes in organizational identity are intimately related to the web of inter-organizational relationships involved in GSW that form a valuable input in understanding their evolution.

Individual-level transformations are significant in GSW as software developers are expected rapidly to switch between different projects, technologies and countries. A large software firm in India will typically have projects in North America, Europe and East Asia, and developers are expected to move between these very different technical, social and cultural experiences. Such radical and constant changes in cultural and work contexts are associated with significant transformations in self-identities. The issue of identity is also consciously adopted by organizations as a strategy in a more instrumental mode, for example to project a certain image for marketing purposes or to facilitate more effective HR management. As organizations increasingly experience the problems of attracting and retaining developers, they will need actively to understand the nature of *self-identity of developers*, and what makes them come to and stay in an organization.

Transformation of identities is a defining feature of contemporary life. In GSAs, the notion of identity is intricately linked with the culture that provides the context within which identity is constructed and with the image that reflects the manner in which identity is presented to the world at large. We discuss the linkage between culture, identity and image in the ComSoft case in chapter 5 to analyse how individual and organizational identifies are intertwined and the manner in which the process of transformation of identity is intricately linked with the growth of the GSA.

Issues of power and control

The manner in which the 'power of flows' dominates the 'flows of power' (Castells 1996) is a key feature of the network society, implying that power will increasingly not be concentrated in institutions but be shaped by the *information flows* that are dispersed over different kinds of networks. Power is provided by the codes of information and images of their representation, and the ability that firms have to access and use that information. In GSW, these codes of information and images are inscribed in systems development methodologies and quality-level certification, and the possession of these codes becomes an important source of power. Whether the 'power of flows' supersedes the 'flows of power' in practice is a complex question, since power relations are constantly being negotiated and redefined over time. In the initial stages of a GSA relationship, the domain knowledge of what is to be developed rests primarily with the outsourcing company and as a result they control the 'flows of power'. However, as knowledge is gradually transferred to the contracting company, there can be a degree of re-configuring of the power differential. The flows of information and knowledge over the network contribute to this reconfiguration, implying that the 'power of flows' can supersede the 'flows of power'.

Setting up GSA relationships is a complex task requiring significant investments in infrastructure. The organization making these investments is in a position of power as they define the standards of work and how they should be enforced in practice. While technologies can potentially be used to redefine the contours of work in terms of the time and space conditions within which activities are carried out, it is still the groups in power that define how this reconfiguring takes place in practice. For example, in setting up schedules for videoconference meetings between groups in different parts of the world, it is often the group in power that defines when the meeting takes place – normally, at a time which is more convenient for one group (middle of the day, for example) which may be late at night for the other group.

In the kind of knowledge work that is associated with GSW, the role of individuals is significant, implying that the power balance revolves around key individuals. The power is often exercised by the management providing extra compensation and attention to these key individuals. In the buyers' market of the global software domain, such power is magnified. Beck (1992) argues that individual power is, however, shaped within a capitalist logic and framework, which is in the control of large organizations. Companies may need developers to work on proprietary languages, for example, which in a sense blocks them from speaking to other companies that use alternative development platforms. The methodologies and processes used in the software development process themselves serve as instruments of power and control, raising the empirical questions of which methodologies are used, who is imposing them and what possibilities exist to use alternative approaches.

The issue of power and control is highlighted in chapter 8 through the Gowing–Eron GSA case analysis.

Cross-cultural communication challenges

Research in international business has established the importance of cross-cultural challenges of communication especially in face-to-face settings. Cross-cultural communication, in general, is hampered by a multiplicity of challenges arising from the difficulties in understanding the 'other person's point of view'. Culture is often the root of communication challenges as it shapes how we approach problems and participate in groups and communities. However, these issues take on a different form and level of complexity when looked at within the context of the temporal and spatial conditions of separation that are inherent in GSW. Communication challenges are magnified by the technical nature of language that is required in the conduct of software development work owing to the multiplicity of products, technical processes, tools, notations and methodologies involved. The use of different and often rather complex forms of ICTs, each with its own conditions and level of reliability in capacity to support interactions between distributed teams, makes cross-cultural communication a central challenge in the GSW context.

Culture is a complex topic, with many different definitions. We use the term 'culture' to refer to a group or community that share some common experiences that shape the way in which its members understand the world. As discussed above in the case of identity, culture is linked with concepts of the identity and image of the organization. Culture is not taken as something 'given', but as something that is continuously constructed and achieved through a variety of complex organizational processes, including that of communication. Communication takes place within different members of the 'community of practice' engaged in the GSA including the tasks of planning and implementing software development activities. These work practices refer to the various socio-cultural processes inherent in the process of knowledge transfer, including the manner in which members draw upon and apply different forms of explicit–implicit, formal–informal knowledge. Of interest here are the processes of communication involved as members of these communities deal with the task of *exchanging information and knowledge* about various products, processes and practices. More specifically, we focus on understanding the various cross-cultural aspects that shape the processes of communication and with it the nature of social relationships between actors that further inform and shape subsequent communication. The nature of relationships depends upon and also reflects the ICTs being deployed (or not), the bandwidth in place and the manner in which they both facilitate and constrain communication processes. The ICT infrastructure in place is a significant determinant to the question of how the balance between the staff situated onsite (in the UK, for instance) and offshore (in India) is developed, which in turn influences the nature of relationships between the various actors involved.

Although ICTs provide the technical capability to develop communication systems that are 'real time', and 'simultaneous', different members of a community of practice (say the Indians and Canadians engaged in a GSA) may have varying norms and assumptions about what 'real time' and 'simultaneity' mean. There will also be differences in the resources and capabilities to access and use the ICTs. People and groups from different cultural backgrounds use different spoken languages and have varying approaches to decision making, conflict resolution, the use of evidence to justify statements, forms of non-verbal behaviour and preferences for written, verbal or graphical communication. The ICTs themselves have varying effects on communication since they favour the preservation (over time) and dissemination (across space) of some kinds of information over others. ICTs also differ in the extent to which they allow immediate or delayed reciprocity, and static or dynamic information exchange. Some organizations thus prefer to use videoconferences for project monitoring meetings, where questions raised can be immediately answered. While immediate reciprocity is an advantage of videoconferencing it is expensive to install and operate as compared to email, which provides relatively 'delayed reciprocity'.

Although, in general, ICTs technically provide the capability to communicate, groups have different capabilities to convert this 'communication' into 'collaboration'. This important distinction is often missed when viewed from an information-processing perspective, which equates communication with collaboration. Just as increased information sharing is a necessary but not sufficient condition for building knowledge, communication and collaboration in a GSA are necessary but insufficient. For collaboration to take place, communication needs to be effectively assimilated and applied in practice to the everyday conduct of the GSA. The influence of communications on the relationship needs to be reflected upon and revised to change behaviour. Collaboration, which can be conceptualized as 'effective communication', takes place under conditions of *mutuality* when members talk 'with' and not 'past' each other (Couch 1989), and this mutuality is directed towards *substantive* content relating to the attainment of project goals.

A variety of technical, social and organizational issues in the context of globalization thus combines to create various cross-cultural challenges to communication that shape the nature of a GSA relationship and how it evolves over time. In chapter 9 we discuss some of these cross-cultural communication challenges in the relationships between Japanese and Indian firms.

2.5 Empirical programme of research

To help answer the question of 'how do GSA relationships evolve over time?' we have discussed six themes that provide the lenses for micro-level analysis of GSA relationships.

Before going into detail of how the cases that we have researched relate to particular themes, we provide an overview of the empirical programme that included these cases:
We investigated three relationships:

- A Canadian company, GlobTel, in GSA relationships with four Indian companies, MCI, Witech, SRS and Softsys.
- A UK firm, Gowing, in a GSA relationship with an Indian firm, Eron.
- A UK firm, Sierra, in a GSA with its subsidiary in India.
- In Japan and Korea we did not investigate any particular relationship but interviewed a number of managers from Korean and Japanese companies who were at a relatively early stage of trying to establish relationships with Indian firms.

The analysis of these three GSAs is presented in seven chapters:

- **Chapter 3**: an overview of the GlobTel GSA programme, extending over their four relationships in India.
- **Chapter 4**: case study of GlobTel–Witech, analysed under the theme of 'standardization'.
- **Chapter 5**: case study of GlobTel–ComSoft, analysed under the theme of 'identity'.
- **Chapter 6**: case study of GlobTel–MCI, analysed under the theme of 'space and place'.
- **Chapter 7**: case study of Sierra UK–Sierra India, analysed under the theme of 'knowledge transfer'.
- **Chapter 8**: case study of Gowing–Eron, analysed under the theme of 'power and control'.
- **Chapter 9**: an exploratory analysis of the relationships between Japanese and Indian firms, drawing upon the issue of cross-cultural communication challenges.

Syntheses of the theoretical and managerial implications that arise from these case studies are presented in chapters 10 and 11, respectively. These syntheses are based upon an inter-case comparison of various theoretical and managerial issues.

A note on methodology

The cases described in this book are part of an empirical research programme on GSA relationships in India. The names of all the companies and individuals mentioned in the description and analysis of the cases are pseudonyms. We have taken care to present the details in such a way that the identities of the entities are disguised as much as possible. The focus of the analysis constantly has been on issues rather than individuals.

The research started in 1996 with a five-person research team: three from a Canadian university and two from an Indian management institute. Over time, two of the Canadians and one Indian researcher left the project, and a UK-based researcher joined it. One member from the original team was present in all the interviews. We thereby ensured continuity in the interpretation of the data. The research programme started with the study of GlobTel, a large North American telecommunications firm, which had established relationships with four Indian companies in the early 1990s. Subsequently, the research program was extended to some UK, Korean and Japanese companies who were at various stages of outsourcing relationships in India.

Table 2.2 Summary of interviews conducted

GSA relationship	Indian respondents	Alliance respondents	Total respondents
GlobTel		25	25
MCI	12		12
SRS	18		18
Witech	15		15
Gowing–Eron	22	19	41
Sierra	14	7	21
Japan	30	25	55
Korea	15	16	31
Total	**126**	**92**	**218**

Two aspects of our empirical program are salient. First, the underlying basis for the empirical programme has been our quest to obtain interpretations from managers and developers on both sides about the state of the relationship, the different challenges and how the respective organizations are dealing with them. The balance between the viewpoints obtained from either side, however, necessarily varied in the different cases owing to issues of access, cost and time constraints. Secondly, a longitudinal perspective was adopted in an attempt to try and understand the issues, the responses and future expectations, on a continuous basis. We tried to visit the sites at least once a year, to conduct repeat interviews at least with some of the key respondents and to get them to reflect on some of the issues that they had raised in earlier meetings.

As can be imagined, conducting such a large-scale programme of empirical research spread over not only multiple countries and organizations but also over time is an extremely complex endeavour. We tried to conduct the study through a network structure of researchers affiliated to different institutions spread over different countries and having close linkages with some of the organizations that were studied. Using this network structure, we conducted more than 200 interviews spread over ten organizations and seven countries in 1995–2000, the details of which are summarized in table 2.2. The analysis of each relationship is anchored by the overall set of interviews, and thus the understanding of a particular case cannot be separated from the larger study of which it is a part. The analysis of any sub-set of interviews is inextricably based on our broader understanding of GSA relationships.

Our research approach was aimed at developing a holistic and in-depth understanding of the process of evolution of a GSA relationship over time. We selected a longitudinal research design, except in the East Asian case. By interviewing people involved on both sides of the relationship in their respective contexts, we tried to understand the issues they were facing, the expectations they had and how these were changing over time. We did not have a fixed schedule of questions but rather a set of issues that we felt were important in shaping the relationship. We would start by asking the respondents to comment on these issues, identify new ones and describe how they were dealing with them. Over time, as we met people for a second and third time, there already existed a shared understanding between us and we could start the conversation from where we had left off at the last meeting. We provide below a brief overview of the different pieces of empirical work conducted in the four sets of case studies.

GlobTel and the Indian firms

GlobTel, a large North American-based telecommunications MNC, had relationships with four Indian firms. In chapter 3, we first present the GlobTel case, and this is followed by analysis of three of the four relationships in chapters 4–6. This includes interviews with all the GlobTel respondents in Canada, their UK lab and their Indian expatriate managers. This perspective is supplemented with some secondary data obtained from company websites and newsletters in order to develop some element of a contextual understanding. By first presenting the GlobTel perspective, we avoided repeating a number of the common elements across the three cases, for example the motivation of GlobTel to start relationships in India.

A total of seventy interviews were conducted (with GlobTel personnel as well as those from the four Indian firms), and except for one that took place over videoconference, all were conducted in face-to-face settings in the location of the interviewee's workplace. The videoconference meeting had four of the five researchers present and the hook-up took place between the University in Canada where the four researchers were present[1] and the respondent in his US office. The other face-to-face interviews had two of the researchers present (with one of the members being common across all interviews over time). Each interview lasted for about an hour except the videoconference, which lasted about 2 hours. All the interviews with the Indian respondents and also the seven GlobTel expatriates in India were recorded. Only four of the remaining GlobTel interviews conducted in North America were recorded owing to the reluctance of the interviewees. The reluctance of the others to be recorded could be because of the more formal requirements of non-disclosure arrangements in North America. We found the Indian companies in general to be more comfortable about recording. The recording of the videoconference was flawed, so we made use of our handwritten notes taken on this occasion.

The first set of interviews that took place in GlobTel's North American office in November 1996 helped to reconstruct the relationship historically, and to develop an understanding of the perceived future challenges. Our request to GlobTel for permission to conduct interviews with their partners in India was granted and we visited the Indian partners in February 1997. There were several limitations to this study. In 1998 we visited the North American office but could only get to meet the staff from the Research and Development office and not the line managers since they were either extremely busy or travelling. We acknowledge this to be a limitation of the research as the line

---

[1] Three of the four researchers were based in this particular university in Canada, while the fourth had travelled from India for a scheduled research meeting of the four researchers. The videoconference was scheduled during this period to take advantage of his presence.

managers' views are crucial to a holistic understanding of the process of growth of the relationship. Another limitation of the empirical study was that after 1998 we did not get further access to do interviews in North America because of sensitive changes taking place in the relationship during that period. This limitation was, however, partly offset by the fact that we met with GlobTel's expatriate managers in India and the UK, and learned the nature of issues as from their perspective.

The Gowing–Eron relationship

Interviews in Gowing began in 1998 as a result of an introduction to the Indian company Eron. Forty-two interviews were conducted with the senior and middle management and the development staff, in both Britain and India. Early interviews concentrated on historical reconstruction of the relationship between Gowing and Eron, followed by interviews at different periods with actors at various levels in the organization. After the initial interviews, follow-up interviews would normally start with participants being asked to bring the 'story' up to date with key events in the evolving relationship. We would then explore these issues and discuss themes that were of relevance in previous interviews as well as future challenges. Information on the relevant context at various levels (company performance, industry trends, etc.) was also collected.

Two field trips to Chennai in 1998 and 1999 led to a total of three days of interviews with key participants, including the programming staff. The UK-based researcher undertook all interviews in the UK. All interviews were taped, transcribed and subsequently discussed with one of the Indian researchers. Subsequent trips were made to the UK offices of Eron and Gowing for formal interviewing and the attendance at meetings when requested by the Gowing management. The UK and Indian researchers, who still remain in close contact with Eron, made one subsequent field trip to Mumbai in 2000 to conduct further interviews. A lot of other secondary data including newsletters, press announcements, etc. were also collected to get a broader understanding of the relationship.

The Sierra subsidiary

The first interviews in Sierra, a UK-based software house, took place in 1998 during the setting up of their subsidiary in Bangalore. Subsequent meetings took place in the interviewees' workplace in either the Bangalore office or in London. The UK-based researcher conducted the interviews in India with at least one or sometimes two of the other researchers present, but he conducted the interviews in the UK independently. Most of the interviews were recorded except for those with the India centre manager where handwritten notes were taken and transcribed in full afterwards. The transcripts were made available to all three researchers concerned for the process of analysis. The final set of interviews with the Indian staff planned for early 2000 were cancelled owing to the centre being closed down.

The analysis of the data was done at periods where the researchers met and discussed substantive themes and issues and related the discussion to theoretical concepts. The writing of reports was critical, and two such reports were written, one early in 1999 and the other in 2000 after the India centre was closed. The reports summarized interviews and identified substantive themes, together with our views on how the process could be improved. We felt it important to provide this input since Sierra was a small company with limited resources and they expected something in return for the time spent in interviews with us. Providing these reports helped to maintain access and served as a useful vehicle for data analysis involving discussion with interviewees.

Interviews in Korea and Japan

Unlike the earlier cases, the interviews in Korea and Japan were not focused on any one particular relationship. Rather we spoke to a cross-section of managers and developers from Japan and Korea who were either at an early stage of outsourcing work to India, or were thinking about doing so. So, in general we have an understanding of the perceived challenges and opportunities of doing work in India. Interviews in Korea only took place in the first round in 1998 and the subsequent meetings focused entirely on Japan. The reason for this was primarily logistical. The interviews in Korea took place in two large Korean conglomerates, in the divisions that were coordinating with the software development activities in India. We did a total of thirty individual and group interviews in Japan and half that number in Korea.

Interviews in Japan were conducted over four visits during which various companies were seen. We also visited the Singapore subsidiary of one of the Japanese companies because software development was being outsourced to India through this 'hub'. To address the problem of language, we conducted interviews together with a Japanese speaker. On the Indian side, we conducted twenty-five interviews spread over four Indian firms that we had studied in the GlobTel case. These firms also had Japanese projects and since the researchers had developed a good rapport with them as a result of the prior relationship, access to their Japanese projects was facilitated. Another reason was that we were interested in developing a comparative analysis with the GlobTel case, and by taking the same Indian firms; we could study the difference in approach to the two sets of clients. In the Korean case, we conducted fifteen interviews in two of the Korean software development centres in India.

Process of data analysis

All the interviews were transcribed (if tape-recorded) or written up as detailed field descriptions (if not recorded). The different researchers present in the interview would make up their own notes and impressions and share them with others electronically. Through discussions we tried to develop cross-perspectives on the interpretations. These different notes and discussions provided the basis for a 'first-cut' on the analysis in the form of a report that was submitted to both the partners in the GSA. We tried to prepare and circulate this report within about 2–3 months of having completed a set of interviews, and submitted three reports over the course of the GlobTel study and at least one or two reports in each of the other cases. The reports served a number of purposes. First, they summarized and consolidated our impressions of the relationship and as such provided a basis for subsequent theoretical analysis. Secondly, they became points of communication between the researchers and the respondents and useful in 'continuing the conversation over time'. When starting a new set of interviews, we would refer back to the previous reports and would urge the respondents to comment on their contents and how they felt things had changed with respect to some of the issues mentioned. These discussions on the reports helped us to 'verify' and cross-interrogate our interpretations. Furthermore, even though these reports were not part of an explicit and predetermined 'action research' strategy, they provided useful feedback and helped to continue our research access. They also improved our credibility at the research site and led to management openly discussing strategic issues with us.

Although data collection was done mainly through semi-structured interviewing, additional data was obtained through company websites, newspaper reports and company newsletters. The broad scope of people interviewed provided us with a wide range of interpretations over time. Our research approach reflects the tradition of studies that can be broadly classified as 'interpretive case studies' (Walsham 1993, 1995). There is an increasing body of work in the information systems (IS) literature based on this approach (for example Suchman 1987; Boland and Day 1989; Orlikowski 1992; Walsham and Sahay 1999).

Although the approach to data collection was empirically grounded and 'bottom-up', the data analysis was interpretive and conducted in a holistic and 'top-down' manner. The interpretive analysis, as mentioned earlier, necessarily had the frame of reference of the different interviews we were conducting and was not restricted to the particular case studies reported. We did not attempt to adopt the use of computerized tools such as 'Nudist' as we felt they would be reductionist as they would not be able to take into account these various other interviews and data sources we were engaged in reading and discussing over time, countries and organizations. We decided to rely on our individual and collective abilities to integrate and synthesize the more than 1,000 pages of data, notes and reports. Key to this integration process were periods of intensive discussion between the researchers who came from different academic and cultural backgrounds and countries and introduced multiple perspectives into the analysis. For example, one of the researchers who lived in India had a deep understanding of the Indian software industry and provided various 'place-based' insights that otherwise might have been missed. A senior Canadian researcher had studied JVs for a long time and could provide the group with insights on the working of such arrangements, especially as viewed from the perspective of a large North American MNC.

The researchers met at least twice a year for periods extending to a few weeks in India, Canada, the UK or Norway for detailed discussions on individual interpretations of the transcripts. During these meetings we discussed various issues and related them to relevant theoretical concepts. This helped to clarify and refine the interpretations, to suggest alternative interpretations and to identify further areas for theoretical investigation. The approach thus involved a continuing dialogue between data collected, our interpretations and feedback from the case participants, discussions with colleagues and our continued reading of related literature. The rapid changes taking place in the industry forced us to constantly review of the interpretations and explore theories to develop insights into the processes of change.

BIBLIOGRAPHY

Bartlett, S. C. and Ghoshal, S. (2000). *Transnational Management: Text, Cases and Readings in Cross-Border Management*, Boston: Irwin McGraw-Hill

Beck, U. (1992). *Risk Society: Towards a New Modernity*, London: Sage

Boland, R. J. and Day, W. F. (1989). The experience of system design: a hermeneutic of organizational action, *Scandinavian Journal of Management*, 5, 2, 87–104

Brown, J. S. and Duguid, P. (2000). Organisational learning and communities of practice: towards a unified view of working learning and innovation, *Organization Science*, 2, 40–57

Castells, M. (1996). *The Rise of the Network Society*, Oxford: Blackwell

  (2001). Globalization and identity in the network society, *Prometheus*, March, 4–18

Couch, C. J. (1989). *Social Processes and Relationships*, New York: General Hall

Dirlik, A. (1998). Globalism and the politics of place, *The Society for International Development*, 41, 2, 7–13

Geertz, C. (1973). *The Interpretation of Cultures*, New York: Basic Books

Giddens, A. (1990). *The Consequences of Modernity*, Cambridge: Polity Press

  (1994). Living in a post-traditional society, in U. Beck, A. Giddens and S. Lash, *Reflexive Modernization: Politics, Tradition and Aesthetics in the Modern Social Order*, Cambridge: Polity Press

  (1999). Runaway world, The BBC Reith Lectures, Radio 4, BBC Education, London

Hannerz, U. (1991). *Culture, Globalization and the World System*, London: Macmillan

Hanseth, O. and Braa, K. (2001). Hunting for the treasure at the end of the rainbow: standardizing corporate IT infrastructure, *Computer Supported Collaborative Work*, 10, 262–92

Lave, J. and Wenger, E. (1993). *Situated Learning: Legitimate Peripheral Participation*, New York: Cambridge University Press

Mowshowitz, A. (1994). Virtual organization: a vision of management in the Information age, *The Information Society*, 10, 267–88

Orlikowski, W. (1992). Learning from notes: organizational issues in groupware implementation, *Proceedings of the Conference on Computer Supported Cooperative Work*, ACM, New York, 75–84

Robertson, R. (1992). *Globalization: Social Theory and Global Culture*, London: Sage

Schultze, U. (1998). Investigating the contradictions in knowledge management, in T. Larsen and L. Levine (eds.), *IS Current Issues and Future Changes*, IFIP Working Group, 8.2, Helsinki

Suchman, L. A. (1987). *Plans and Situated Actions, The Problem of Human Machine Communication*, Cambridge: Cambridge University Press van Maanen, J. (1989). Some notes on the importance of writing in organization studies, *Harvard Business School*, Research Colloquium, 27–33

Walsham, G. (1993). *Interpreting Information Systems in Organizations*, Chichester: Wiley

(1995). Interpretive case studies in IT research: nature and method, *European Journal of Information Systems*, 4, 2, 74–81

(2001). *Making a World of Difference: IT in a Global Context*, Chichester: Wiley

Walsham, G. and Sahay, S. (1999). GIS technology for district administration in India: some problems and opportunities, *MIS Quarterly, Special Issue on Intensive Research*, 23, 1, 39–66

3 GlobTel's GSA programme in India

3.1 Introduction

In this chapter, we provide a broad overview of the initiation, expansion and stabilization of GlobTel's GSA programme in India that includes the four relationships with Witech, MCI, ComSoft and SKA (all pseudonyms). Two of these firms (MCI and Witech) were large and established Indian software houses, while ComSoft was a start-up company established by Indian technocrats with Silicon Valley roots, and SKA was a software firm established by a medium-sized business house with limited experience in software development. MCI was based in Mumbai, SKA in Delhi and the other two in Bangalore. The relation with SKA was short-lived owing to perceived business incompatibilities with GlobTel, and after two years Softsys replaced SKA as the fourth partner.

The overview of GlobTel's programme is based broadly on the interviews taken with senior managers of GlobTel in North America, their expatriate managers in India and with senior members of the staff in one of their labs in the UK. Background information has also been obtained from the company Website and newsletters. The overview is at a macro level, encompassing the four relationships, and not restricted to any particular GSA. The overview is at a rather 'factual' level of details of key events in the relationship, including a historical reconstruction of events prior to 1996 when our research started, and then an account of key events as the research progressed from 1996 to 2000. This overview provides the backdrop for chapters 4–6, where three of the individual cases are theoretically analysed. We do not present the Softsys relationship separately, as there were elements in that analysis which we felt overlapped with the Witech case that is described in chapter 4.

We start by providing a brief background about GlobTel, with a focus on the company's R&D activities. This background is not meant to be a comprehensive account of their complex history, that goes back more than 125 years, but to provide some key information which helps to contextualize the Indian relationships. A description of the phases of initiation, expansion, and stabilization of the GSA programme in India is then presented.

3.2 Background

GlobTel is a large North American multinational with a powerful history in which it has moved from the manufacture of telephones, to digital technology, to public and private networks and currently to Internet-based technologies. Some of the new and exciting streams of revenue include optical networks, wireless Internet solutions and intelligent networks.

GlobTel continues to be a very large, powerful, and successful global telecommunications vendor. It is among the top telecommunications equipment manufacturers in the world. In the 1970s GlobTel was primarily a national manufacturer of telephone equipment primarily for the local market. At the time of starting our research in 1996, it had about 68,000 employees, and worldwide revenues of US$12 billion, including US$500 million in profits. In 1996, GlobTel was the sixth largest telecommunications vendor and third highest cellular equipment supplier in the world. In 1996 it had customers in over 150 countries, with about one-third of its employees and half of its R&D operations and 11 per cent of its revenues coming from national operations. Roughly 45 per cent of its revenues came from outside North America, including Europe, Asia and Latin America. In 1996, GlobTel's product lines could be broken into four categories:

- **Public carrier networks** for phone companies across the world
- **Broadband** – cable, TV and new long-distance carriers
- **Wireless** (cellular-based technologies)
- **Enterprise networks** – internal communications for large enterprises.

In 1997, switches accounted for 27 per cent of its revenues, while wireless accounted for 22 per cent. After a period of tremendous churn and upheaval in the telecommunications industry globally, GlobTel, with a new name and focus, has (2000) an annual turnover of more than $30 billion, customers in more than 150 countries, R& D facilities in seventeen countries and multiple manufacturing facilities globally. GlobTel is operating manufacturing and R&D facilities in many different countries throughout North America, Asia Pacific, the Middle East and Europe.

GlobTel is more than 125 years old. It focused on manufacturing until 1960 when it created an independent R&D division. Even though the internal R&D function had started in the 1930s, these efforts were given a big push in the 1960s, and the strength of the R&D personnel expanded from an initial forty-two engineers to 800 in about five years. Around this period, the first international manufacturing was also established in Europe. In the early 1970s, a conscious attempt was made to reinvent the R&D culture by developing an environment akin to that of a graduate university, an environment that is conducive to free thinking and innovation. This strategic thrust in R&D contributed to nearly 75 per cent of sales being based on internal designs compared to less than 10 per cent in 1960. GlobTel developed a series of electronic digital switches, including a pioneering breakthrough of the all-digital central office switch that made it a leader in the industry. The company was renamed in the mid-1970s, which we refer to by the pseudonym of GlobTel.

The period since the late 1980s has been turbulent, with a focus on cost reduction, including a cutback on R&D expenditure. In the early 1990s, there was a new CEO who, despite continuing some of the cost-cutting measures, restored spending on R&D. In the mid-1990s, a policy decision was taken that all the GlobTel labs worldwide would come under a common banner and brand name (GlobTel). Around this time there was another change in the CEO, but the new incumbent continued the direction of his predecessor in providing a strong R&D focus. In the late 1980s, about 45 per cent of GlobTel employees had been classified as 'knowledge workers'; by the mid-1990s this figure was about 70 per cent and the current figure is in excess of 75 per cent. This transformation had been fuelled by the increased expenditure on R&D, to focus on new product and services. A senior GlobTel executive described this transformation process in a public meeting in 1998:

GlobTel has moved from the labor-intensive manufacturing of hardware – of boxes so to speak – to creating the software and communications technology for a wide variety of network solutions and services that require heavy investment in research and knowledge workers. Our future is going to be even more closely linked to the development of the Internet. We can lead the telecom industry from dialtone to what we call the webtone.

The move from dialtone to webtone was part of the new 'right-angle turn' strategy of GlobTel in 1998. At that time, the company had about 73,000 employees and the strategy entailed laying off people and also hiring new ones in the quest to change from a capital-based to a knowledge-based orientation. Telecommunication companies like Lucent were no longer the only competitors to GlobTel, as data companies like Cisco also were now challenging them. This shift in competitive focus reflected the significant upheaval taking place in the ICT industry in the 1990s catalysed by the proliferation of the Internet.

The transition of this giant firm from a 'pure' telecommunications to a 'data oriented' telecommunications form has not been without its problems. In 1998, the company lost about US$11 billion market capitalization, reflecting investors' loss of confidence in the company. Some of GlobTel's acquisition of technology companies also raised questions, and shook customer confidence with a negative effect on the share prices. Our research took place during this extremely turbulent period of transition. In order to make its R&D activities globally competitive and cost effective, GlobTel pursued a strategy of expanding its network of international alliances and partnerships. Globalization of R&D activities would also help GlobTel to get closer to customers and to accommodate different standards, approaches and customer requirements around the world. It was in the early 1990s within a backdrop of rebuilding the focus of international R&D that the software alliances were created with the four Indian GSAs. A key difference of the

Table 3.1 Overview of the issues in GlobTel's externalization programme in India, 1991–2000

Phases of the relationship	Key issues
Initiation (1991–6)	Motivators: cost efficiency considerations, resource crunch, globalization imperatives and using the enabling force of technology
	Global R&D Group (GRDG), the hub of the initiation effort
	Some initial resistance from GlobTel line managers
	Indians have little prior experience of telecommunications or in doing GSA work
Expansion (1996–8)	Projects selected based on potential for offshore work rather than 'body-shopping'
	Steady evolution in the level of work
	Steady growth in infrastructure
	Growth in problems, including attrition and perceived lack of long-term commitment
Stabilization (1998–2000)	GlobTel managers less upbeat on the relationship, recognizing the inherent nature of the problems
	Indians demanding a different pricing basis to their work in tune with the higher level of work they were doing
	GlobTel reluctant to change the pricing basis
	New competition in the industry
	New technology focus in the industry induced by the Internet
	Both sides acknowledge learning through the relationship

Indian alliances as compared to other GlobTel alliances was that they did not cater to local markets and served rather as a resource base for GlobTel's global R&D operations. In the second part of the 1990s, during the period of transition, the four relationships went through frequent changes. It is against this background that we provide an overview of the genesis and growth of GlobTel's GSA programme in India.

3.3 GlobTel's externalization to India

We present a summary of key issues in GlobTel's externalization programme in table 3.1, and then elaborate these issues in the discussion that follows.

Initiation of externalization (1991–1996)

By the end of the 1980s, there were various labour statistics' reports predicting a large and increasing shortfall of developers in North America. These reports were making firms seriously think about strategies and options to meet the software needs that in the meanwhile were growing exponentially. The idea of GlobTel having partnerships in India as a source of software R&D emerged in the late 1980s, in a discussion between Soumitro Ghosh, the Vice President of the Global R&D Group (which we call the GRDG), and the then Chief Operating Officer (COO) of GobTel who later went on to become the CEO. Ghosh, an Indian by birth, had lived in North America for over two decades and harboured a strong desire to contribute to the development of his homeland. The issue of developer shortages in North America triggered the discussion between Ghosh and the COO and the knowledge that other companies like Unisys and Texas Instruments were already doing software development in India. GlobTel was also, of course, no stranger to global operations. The COO, who at that time was heading the wireless division of GlobTel and was an influential figure in the company, thought this idea was worth pursuing. Three senior members of the staff from GRDG including Ghosh, Paul and Shinde, all of Indian origin, conducted an initial feasibility study. While Ghosh and Paul were responsible for nurturing the relationship in later years, Shinde died of a heart attack soon after. Ghosh and Paul played a key role in shaping the nature of evolution of the Indian programme.

Ghosh and Paul were both charismatic personalities with a strong desire to innovate and take risks. For example, Ghosh argued against the traditional model of innovation (starting in high-cost countries and gradually moving to low-cost ones), as being too orderly, predictable and static. This traditional model was questionable because access to foreign markets despite financial backing and advances in communication technologies had enabled global networking in such a way that sharing of ideas did not necessarily need to follow predefined trajectories of innovation. Paul described himself and Ghosh as 'shock absorbers' for the GlobTel management. He believed that their Indian background would enable the programme to be built on a sound understanding of local problems in India. The strong support of the GlobTel management, coupled with their ability and desire to innovate, as well as their individual personalities and charisma, helped Ghosh and Paul to seek out and develop personal relationships with Indian companies.

The 'externalization' programme, as it was known in GlobTel, started with the establishment of a liaison office in the early 1990s following the feasibility study conducted by Ghosh, Paul and Shinde. This program was part of a larger network of GlobTel's international R&D initiatives in Eastern Europe, China, India, Vietnam and Brazil. Right from the start, India was seen as a long-term initiative. GlobTel did not follow the short-term and opportunistic cost-cutting strategy that was typically adopted by many North American firms. Four Indian firms were initially selected as partners, primarily based on their diverse technical strengths and the congruency of their business practices and principles with GlobTel.

At the point of the initiation of the externalization program, private sector involvement in the Indian telecommunications business was relatively limited, and expertise was primarily located within public sector units like the Indian Telephone Industries and the Centre for Development of Telecommunications. The four Indian firms, like most in the industry, had limited prior expertise in telecommunications. GlobTel's initial motivation for externalization was primarily a resource issue, although they saw a future potential to tap into the Indian and possibly also the Asian markets. Speaking at an international conference, Ghosh described GlobTel's motivation to outsource to India to be based on four reasons: cost efficiency considerations; imperatives of globalization; coping with the ongoing resource crunch; and the enabling potential of technology.

- *Cost efficiency*: Ghosh said:

  Offshore contracting to a country such as India can ease the R&D budget crunch since its lower cost of software development can create a powerful multiplier. For example, take the case of proposed software development that costs $100 but available funds amount to only $75 leaving a shortfall of $25. Taking advantage of an offshore multiplier of three, a US company could divert $10 of its funding offshore to produce as much software as it would take $30 to produce in North America. At that point the budget of $75 would have purchased $95 worth of software and a shortfall of only $5 would remain. If the company diverts $12, there will be no shortfall.

- *Imperatives of globalization*: Ghosh said:

  North American corporations can become globally more competitive if they can develop software more creatively, more efficiently, and more economically. Having an international software operations affiliate is not just a cost saving measure. It can be a strategic move designed to provide a competitive advantage over more centralized, narrowly focused competitors. Indeed, it should be the objective of any growing corporation to transform into an international and multi-talented organization. Apart from GlobTel, many other companies are already active in India including Hewlett Packard, Texas Instruments and British Telecom. Can any North American company (irrespective of size) afford to forgo the competitive benefits that such companies are realizing from such international software operations?

- *Coping with the resource crunch*: Ghosh argued that India, with its strong educational infrastructure and English-language capabilities provided GlobTel with a rich source to access software developers. Another GlobTel senior manager said 'on the top of the list [of reasons] was that there were not enough people in North America to hire on, to train to do our project work. So we wanted to externalize and get other people throughout the world to do development.' This access would allow GlobTel to free their domestic resources to concentrate on activities closer to their customers in North America, and potentially roll out new products with shorter time to market. Externalization, Ghosh argued, provides various benefits including workload balancing, shorter time to market and round-the-clock software development. It would, in addition, help develop strategic partnerships with Indian institutions that could open up potential market opportunities in India.

- *The enabling potential of technology*: Ghosh said:

  One issue of special note in our experience in India is the way in which GlobTel uses telecommunications to make the partnerships work. We enhance our ability to communicate with our partners by investing in and engineering data, voice, and video links to their offices. By connecting them directly into our global, corporate-wide area network, we reduce the influence of distance. Boston and Bombay are only three seconds apart. In addition, we replicate our software development environment at the partner's location. Today, we develop software in India at a distance of 10,000 miles as an integral part of our global business process.

Despite the intention of establishing long-term relationships with Indian partners, GlobTel opted for vendor–contract relationships with all the four partners rather than

JVs. However, the HR Director at GRDG described this as a 'services agreement' that in practice was more like a 'joint venture'. Another GlobTel expatriate manager said, 'it was possible to have a legally sound, binding commitment that may be a customer-contract relationship on paper, but at the working level, it could be a partnership'. In this way, GlobTel attempted to get the best of both worlds – the benefits of a JV without necessarily exposing itself to the risks that such a partnership entailed, for example through financial investments. However, GRDG was open to the possibility of the relationship evolving and taking on a different form in the future. An Indian-based GRDG Director justified the choice on the ground that: 'right now GlobTel has not been doing well in selling in India so we are not looking for a joint venture. If we were to enter the market in a big way later, we can then think of changing the relationship with the partners [to a joint venture].' Another senior GRDG Director said: 'I am trying to convince GlobTel people to make a joint venture of some of these relationships, and not an arm's length contractor relationship.'

The work externalized from GlobTel to India initially was relatively routine, involving bug-fixing for software of telephone exchanges. Telephone exchanges have historically been a major revenue earner for GlobTel. The hardware of these exchanges involves standard processors. Different kinds of functionality in the operation of these exchanges, which constitute the competitive strength of the company's products, are provided by custom-developed software. This software for the older exchanges has been developed over a period of twenty years and constitutes more than 24 million lines of code. A senior GlobTel executive told us this software library is second in volume only to NASA. A major part of this software had been coded in a proprietary language (which we refer to as GlobSoft). The systems with the underlying software constitute a stable and reliable product that continues to be used extensively by large telecommunication service providers and also new customers all over the world. In this scenario, the maintenance of software modules which have been developed in response to market changes over two decades is a critical task from GlobTel's perspective. However, this work is not very interesting to programmers who generally prefer to work with newer technologies and non-proprietary or 'open' platforms. Maintenance work is also seen as a cost that needs to be minimized in order to further support the development of new technologies. Reducing costs in an environment characterized by increasingly rising salaries of development personnel is extremely problematic, making maintenance of legacy systems a prime candidate for outsourcing.

Actors involved in the externalization network

The GRDG was the hub in the network for externalization, playing a key role in building multiple relationships including those between the GlobTel line managers and their Indian partners, and between the partners themselves. One Director described GRDG's role as similar to that of a broker or real estate agent, who had the responsibility of matching the needs of line managers with partner strengths within a broader framework of externalization. The initial criteria for matching were the urgent need to satisfy R&D resources and carry out complex R&D projects in cost effective ways. Within GlobTel, the budget structure for R&D had changed from one of a centralized pot of money to a divisional allocation based on lines of business. As a result, a GRDG Director commented, 'now the money on R&D is extremely visible, and dependent on lines of business under the direct control of the business manager. So, externalization will be based more on cost rather than on urgency in the future.'

While the 'champions' within GRDG were promoting the idea of externalization, there was an initial reluctance among some of the GlobTel line managers owing to their lack of confidence in the ability of Indian partners to deliver the necessary outcomes and their discomfort in managing work at a distance, particularly when that work involved different cultural conditions. In some quarters, there was also a fear of job losses in the parent labs if the Indian experiments were successful and work was increasingly shipped out to them. One line manager who was unconvinced about the virtues of outsourcing to India said, 'I don't think having development spread across multiple sites and time zones is conducive to delivering things fast with a high degree of quality. While cycle time can be reduced, fragmentation of other activity hurts quality.' He did not think he would have taken this route if it was his company, and was only following orders. He thought some of the initial equipment was sent to India only because it was not used in North America. They could thus go along with the top management decision to externalize, and yet not part with anything important. Not all the views were negative, however, and some managers saw the potential for positive effects. The following quote summarizes the views of the positive thinkers:

Having a remote lab that is almost 12 hours out of phase with the parent lab allows two days of work to be done in one day. They can build when I go home at night and can be ready for us when we return next morning.

Within this initial scenario of uncertainty about what to expect in India and the pockets of internal resistance, GRDG played an aggressive role, which it was expected (and hoped) would scale down over a period of time with the fostering of more direct contact between the line managers and development partners. At the Indian end, GRDG was seeking to foster a sense of collaboration between the four development partners, a non-trivial task since traditionally these firms had seen themselves as competitors. Overall, GRDG had a very complex and important role to play in the initial stages of the program in building multiple relationships, lobbying with the Indian government for various permissions, creating communications and software development infrastructure, nurturing confidence and trying to develop a shared understanding about the aims and means of the externalization programme among the different actors involved.

Since the relationship was not a JV, it did not share equity. Nonetheless, there was still substantial sharing of resources and risks by both parties. GlobTel made significant technical and management investments. On the technical side, investments were made in infrastructure including replicated servers, high-speed links to facilitate easier partner access to software libraries and better utilization of software tools (for example, being able to access one page at a time rather than one line at a time). Improved infrastructure would enable more complex projects involving interdependent communication between North America and India. Significant management investments were also made in the field of training, especially in the technical domain of telecommunications. A senior GlobTel manager described the initial projects given to the Indians as a 'training mode where you are not trying to pay them for results, but when you have reached a certain plateau, you need to come up with the next step function'. In addition to technical training, efforts were made to support interaction between the personnel from both sides through specially designed orientation programmes for Indians visiting North America on local customs and culture. At the initial stages, such reciprocal programs for North Americans going to India did not exist.

Expansion of externalization (1996–1998)

GRDG initially selected pilot projects from the line managers and allocated them to the development partners. A GRDG Director described the selection criteria for the projects as follows:

They [the GlobTel line managers] ask, 'Can you get me some ten people?' For what, we ask? 'Anything that smells of body-shopping, we reject it. If there is continuity in the project, or potential of continuity, we may support.'

The above quote is significant for at least two reasons. One, it shows the key role of GRDG in the initial selection of the projects. Two, it emphasizes GlobTel's commitment to support only those projects that were to be genuinely developed offshore from India and to reject 'body-shopping' activities. Initial projects were simple jobs requiring minimum direct interaction of the Indians with GlobTel. A GlobTel expatriate described the initial projects to be related to mature software, which also helped to minimize the internal perception that GlobTel were losing expertise to an external agency. As the Indians developed technical capabilities and core competencies in telecommunications, they moved up the 'trust curve' and took on more complex projects that required greater interdependent team processes. This movement in trust levels was evident in the attitude of a GlobTel line manager who, despite being sceptical initially about the Indian relationship, said: 'We could have written contracts with very strict clauses. But I like to deal with them the way I deal with any other group. In the year before they demonstrated their capability and I trust them now.' He added that this trend would continue in the future:

I am trying to give them high visibility feature, a little more challenging work. I am involving them in the development of a new platform so that they can retain more people. I will grow faster in India than I can in North America.

A key issue initially, especially from GlobTel's perspective, was dealing with the uncertainties of distance. The HR Director at GRDG described three strategies adopted by line managers to deal with the uncertainties of distance:

When we start a project we can do things three different ways. Our first preference is that a project leader from GlobTel goes to India and meets the project leader there and his team to explain things. The second alternative is that the Indian project manager comes to North America for a discussion. The third is, no one goes and they have a videoconference. But this is not the same as meeting face to face. We really push them to go over.

A senior GlobTel line manager based on his prior experience of GSA in Israel had abstracted a three-year model of 'technology transfer' to India. This included the first half of the initial year on establishing infrastructure including captive offices, telecommunication links and the software development environment, and in the second half to transfer the software to India. The following two years were spent in the more complex task of the transfer and strengthening of various project management practices. GlobTel adopted an interesting mix of face-to-face and electronically mediated project management. For example while expatriates provided face-to-face control, electronic mediation took place by shifting servers to India. This allowed local login rather than remote access to the servers from India where fast connection was typically obtained in the night time. With time, trust gradually grew, and the Indians, who were initially resistant to the presence of GlobTel expatriates in their offices, started to accept them and acknowledge them as a useful help to resolve technical problems. The investments made by GlobTel in the relationship – for example, in servers and library systems – were seen positively by the Indians as a sign of commitment to a long-term business. The Indian partners also reciprocated the investments by establishing independent laboratories, allocating workstations and providing training.

Despite an initial lack of expertise in telecommunications, GlobTel acknowledged that the Indians were willing to learn, and as a result could move from simple jobs to more complex projects rather quickly. It was also felt that the Indians had the potential to take on ownership responsibilities in the longer run. They did not, however, gain intellectual property (IP) rights (which GlobTel was unwilling to relinquish). The Indians were keen to learn new technologies and take full control over features and sub-products, and they harboured expectations of having products of their own. The HR Director, however, captured the expectations of GlobTel, in this statement:

We would like to evolve to a stage where they take up ownership of a product. People call them whenever there is a problem. They do the upgrades. They do the enhancements, the evolution. Intellectual property is the only thing we will own and there is absolutely no way that we could give that away.

Although legally the relationships continued as contractual vendor relationships, GlobTel wanted the Indians to see themselves as 'virtual GlobTel labs'. A GlobTel Director in India said 'even though we have four partners, they are all virtual GlobTel labs. You are all GlobTel in here, and not Witech, not Softsys, not ComSoft and not MCI. You are all GlobTel.' GlobTel tried to implement this thinking operationally by encouraging the Indian labs to interact with each other as GlobTel labs globally would. This implied a changing mindset in which the development partners would view each other not as competitors but as partners and collaborators. They would share information about projects, technologies, training and satisfaction levels. This identity of a 'virtual lab' where the Indians were partners and not contractors was not easy to implement in practice, as it would necessitate the Indians gaining more access to confidential information about technologies and pricing models. This information GlobTel was not always ready to share. The Indians, too, did not feel comfortable in taking on a new role, as a GlobTel expatriate remarked:

The Indians see themselves as providing service to a client. Even though we try to stress that this is a partnership as opposed to a client–service provider or client–contractor relationship, they still do not feel comfortable demanding some support because you cannot do that to a client.

Technology was the key to the development of the relationship. Communication links were critical, and breakdowns led to significant drops in productivity. A GRDG Director said: 'An issue is the umbilical cord [communication links] which ties them to us. If a lab in India loses communication links with us, there is zero productivity. India right now has a Department of Telecommunications strike, and we lost all links. There are long periods of low productivity.' Videoconferencing played an important role in project management, both to exchange and coordinate information and also to develop trust and confidence. However, technology on its own was not enough, as there was a lot of logistic support required in setting up the times and booking rooms for the meeting. Both sides needed to make adjustments to their work schedules to attend the meeting, and often the Indians felt that their comfort levels were sacrificed for GlobTel in the process of scheduling meetings. A GlobTel manager described the Indians' viewpoint:

There is some resentment. It is very difficult because to communicate with India you have to deal with a $10\frac{1}{2}$ hours time difference. You have to stay up late so that you can talk to them. If there is resentment, it is easy not to answer the phone, or not get back the email. We are starting to hit a lot of bumps around the soft skills. You have to be willing to make that phone call from your house at eleven in the night or come to the office at six in the morning to get through. There are sacrifices that you must be willing to do.

A number of other issues were becoming a cause for discussion and concern among the partners, especially from the perspective of GlobTel, including attrition and measures being taken by the Indian partners to prevent it. While GlobTel thought the measures were not enough, the Indians saw some of GlobTel's efforts as interference into their internal affairs. Another issue related to the nature of long-term and large investments made in telecommunication switches that required 'evergreen' maintenance. As a GlobTel Director commented, the Indian developers with their computer science orientation were reluctant to make such long-term commitments:

I cannot go back to the customer and say things have changed. So, the evergreen concept is that whatever investments are done in the past, are valid today. The platform is green for extended periods, ten, fifteen or twenty years. Equipment that you have bought becomes obsolete. Each machine is worth some four million dollars, once he invests, it stays, and he expects you to maintain it. This is not the norm of the computer industry; if it is obsolete you change it.

Within this context of growth in the level and volume of growth, and associated management tensions, the stabilization of the relationship is described.

Stabilization of externalization (1998–2000)

By 1998, the relationship through joint efforts of the partners had evolved and reached a state of maturity wherein each side was exploring various options by which further value could be added to their respective businesses. The maturing of the relationship introduced new dimensions and complexities, and also created further challenges and opportunities. By the middle of 1998, the GlobTel managers seemed less upbeat about the Indian partnerships than when we first met them in 1996. In 1996, negatives in the relationship were considered teething problems which are likely in the early stages and which could be addressed with more effort. However, by 1998, there was a stronger feeling that these problems were inherent in the process, enduring, and difficult to solve. From the GlobTel perspective, some of these problems included the inability of the Indian managers to deal with attrition, inadequate progress in the relationship, lack of proactivity of Indian managers, and significant problems in the absorption of technology. Maturation also raised different expectations as the Indians needed to take on ownership and marketing responsibilities, and GlobTel managers needed to 'let go' and allow the Indians to manage independently. The Indians needed higher levels of management expertise to deal with ownership and provide service to global customers at the same quality level that GlobTel had previously accorded.

Another challenge arising from maturation and rising level of work was felt on the pricing of the existing vendor–contract relationship that had been based on time and material. As the Indians took on more ownership responsibilities, they wanted to be paid not on the basis of the time they spent on the project, but on the value they brought into the relationship. As GlobTel was reluctant to change this model, it raised a debate about the need for alternative organizational arrangements. While the Indians felt the present contractual form would be difficult to implement in the more unstructured tasks that ownership entailed, GlobTel thought the adoption of alternative organizational forms like joint ventures was premature. The reasons were the variety of complex issues it introduced such as intellectual property concerns, and the changing face of technology and competition in the industry.

The entry of firms including Lucent, Cisco, and GE into the Indian telecommunications market put pressure on both the retention of people and the pricing standards. The Internet was catalysing changes in competition through the convergence of voice and data technologies. As a result, GlobTel embarked on a 'right-angle turn' strategy to become a 'voice over data' rather than the 'voice' company it traditionally had been. The CEO urged that the preconditions of this shift were 'trust, agility, and alignment of goals, and most important of all, a real market focus.' This contributed to rather dramatic changes within GlobTel including layoffs and relocations of existing staff, closures of existing projects, and some bold new acquisitions. For example, in 1998, GlobTel in a US$9.1 billion deal acquired a US high-technology firm to create a partnership 'suited to this era of end-to-end mission critical IP networks.' The strategy of new technology acquisitions rather than in-house R&D investments raised questions about the future of the Indian relationships, and what projects would be externalized in the future. This uncertainty and churn made GlobTel, who otherwise might have been more patient with the Indian partners, to be much more critical. The company became concerned primarily with its own existence.

Despite this ongoing churn and problems, there was no doubt that the relationships had been a real learning experience and had survived the test of time. The relationships were still going strong after nearly a decade that was longer than the average duration of a typical joint venture. Both sides had learnt significantly from this experience, though in different domains, the Indians in telecommunications and GSA relationships, and GlobTel in doing business in India. Paul, one of the champions of the Indian program reflecting on the benefits from the Indian experience said:

One of the things that we can be rightly proud of is that we put in very little money up front. We kind of grew the thing organically. Every time it was giving us back more than we had to put back into it, and that itself is an achievement. GlobTel could grow the labs with minimum risk.

The overview of issues at various stages of the GSA program provides the background against which three of the individual relationships are analysed in micro-level detail in three subsequent chapters. The aim of this analysis is to understand how particular GSA relationships evolve over time and is guided by different theoretical perspectives. These theoretical perspectives have been formulated based on an 'ongoing conversation' between the empirical analyses of particular cases and theoretical understanding developed from our readings on globalization.

4 The GlobTel–Witech relationship: a 'standardization' perspective

4.1 Significance of standardization

In this chapter, we analyse the GlobTel–Witech GSA relationship using the perspective of standardization, which refers to the socio-technical processes through which various standards are implemented in organizations. The standardization perspective is significant since many firms with global operations base their strategy on the assumption that there are more similarities than differences in the world. Coordination of these global activities is then carried out through central control from headquarters, based on standardized products, processes and practices. Standardization becomes a key feature to build and sustain a relationship by homogenizing operations to the extent that the outsourcing and outsourced firms cannot be distinguished from each other. In software development, the role of standardization has in the past been discussed in the context of 'internationalization' of software packages (Taylor 1992; O'Donnel 1994). The purpose of internationalization is to develop and market packages in a 'mass-production' mode. The analysis of standards has not been conducted in the context of the processes by which GSW is carried out. Also, the role of standards in the internationalization of software has largely assumed that through appropriate language-translation strategies, effective global products can be developed and cultural differences taken into account. For example, Taylor (1992) writes:

> The end goals of internationalisation, then, are to be able to have a sort of generic package, with an appendix or attachment that details all the cultural specifications. (1992: 29)

Standards serve the purpose of being a *reference* (for example, weights and measures), or developing *compatibility* between different systems (for example, plug and socket), or specifying *minimum acceptable* levels (of software quality for example). The idea of a standard is linked closely to the notion of a 'universal', implying that one benchmark can apply to all activities and actors within a particular domain. So, a communication standard implies that all actors using a particular technology for communication would be subject to standard language protocols for communication. The idea of standards is not new, and was an important aspect of Adam Smith's notions of routinization spelled out in 1776. This could split even the manufacture of a pin into different tasks allocated to workers based on skills required. Instead of employing highly skilled and expensive workers to do the entire job from start to finish, sub-tasks could be split up among less qualified workers. In the early years of the twentieth century Fredrick Taylor's scientific management principles, based on similar concepts of routinization, increased the emphasis on productivity measurements of different sub-tasks based on predefined standards. While these principles of routinization and standardization have been widely implemented in the manufacturing sector, in recent years they have also been applied to the analysis of the service sector. In a wonderful ethnography, Leidner (1991) describes at length the processes of standardization as they play out in 'interactive service work'. While the objectives of standardization to support mass production are common to both manufacturing and service work, Leidner emphasizes the importance of distinguishing the processes through which standardization is achieved in the two cases.

Standardization has traditionally been an important issue in organizations. Enabled by processes of globalization, some organizations have become too large and diversified for tight central control. They have simultaneously become increasingly embedded in different contexts where they need to understand local particularities in a 'deep' and systematic way. This issue is especially relevant in the software industry where firms operating in a network structure seek to develop shared templates within which they can interact, share information and communicate with others in globally diffused networks. Today, both large and small firms are establishing GSAs with global partners to enable distributed software development. A number of countries and firms are involved in the network with design taking place in one country, development in another and integration and testing in a third. Typically, one location serves as the hub and is responsible for coordinating the different pieces of software development occurring in the various nodes of the network. Coordination is enabled by a networked technological infrastructure and the use of standard product designs, development methodologies and benchmarked management processes that serve as 'best practices'.

The various pieces of technological and managerial infrastructure that underlie a GSA, are held together by various standards – formal and informal, explicit and im-plicit – and represent expert processes that Knor-Cetina (1999) describes as being characteristic of 'knowledge societies'. She emphasizes the understanding of the func-tioning of these processes as largely an 'empty space', that needs to be charted through empirical investigations. Through this case, we seek to chart some of these contours in the GSW domain, specifically focusing on the role of standards in sustaining, en-abling and constraining the evolution of a GSA relationship. Knowledge required for the functioning of GSAs is not merely an external intellectual or technological product, but a production context that is developed over time and comprises heterogeneous elements bound together in a widely extended network. GSAs are fundamentally facili-tated by complex socio-technical 'information infrastructures' (Hanseth 2001) includ-ing high-bandwidth telecommunications links, management practices and procedures and software development methodologies and practices. This infrastructure is sustained through a shared understanding in both the technical and management domains about how software development should go on, reflected in the existence and use of various standards.

To take into consideration these various facets of the GSA relationship, we adopt a general and inclusive conception of a standard implying a *simplification and abstraction with the aim to define and communicate significant aspects of the processes, artefacts and structures across time and space. The objective of all standardization processes is to enable some form of universalization and mass production.* In general, standards are agreed-upon rules for the production of (textual or material) objects required because they span multiple communities of spatially distributed practice (Bowker and Star 1999). Although standards help to provide a sense of stability for those involved in using the infrastructure on a day-to-day basis, as their temporal and spatial scope increases they take on an increasingly inertial nature and create an 'installed base' that makes it difficult and expensive to change. Unlike many other domains where standards are defined and enforced by external agencies (like the World Health Organization (WHO) for health standards), in GSA relationships standards are negotiated 'internally' at the social, political and cultural levels by the parties involved. This processes by which standards are negotiated, implemented and redefined is what we refer to as 'standardization'. Our interest in standards is in understanding their role in *shaping the process of evolution of GSAs* and it extends beyond the technical concerns of individual systems or the protocols to include the relationship in its totality and the standards that underlie it. These include standards for technical and physical artefacts, software development processes, management practices and the processes by which they play out in the everyday conduct of GSA relationships.

4.2 Case narrative

Witech Systems is a part of Witech Infotech that was set up in 1981 as a key business within Witech Limited, a large Indian conglomerate established in 1947. Witech Limited deals with an extremely diversified set of products and services including oils, soaps, toiletries, lighting products, medical systems, fluid power, pneumatic cylinders, financial services and computer hardware and software. In 1996, the Witech group of companies overall had a turnover of US$450 million. Witech Infotech was the biggest contributor to this turnover, with US$300 million. Witech Infotech is primarily in the computer business, focusing on hardware and software. It was the second-ranked computer company in India in 1987, second in software and first in peripherals. Prior to 1988, Witech Systems was catering nearly 100 per cent to the domestic market. However, since the domestic market was seen to have a much slower growth rate than global markets, Witech Infotech, like many other Indian software firms, started to internationalize and entered into a number of tie-ups with MNCs for both the manufacture and marketing of computer peripherals. A conscious decision was taken to try and move away from products to system integration solutions.

On the software side, Witech Systems (which we refer to as WS) included three business units including one in telecommunications. It was this business group that entered into the GSA relationship with GlobTel in 1991 and is the focus of this case analysis. WS focused on systems integration projects for an international clientèle. It was described by the Group Vice President as being 'very strategic and our whole management watches what we do very closely'. The WS–GlobTel relationship is described in three phases of initiation, growth and acceptance of the legacy route.

Initiation (1991–1996)

The relationship was initiated in 1991. Recollecting it about eight years later, participants have varying opinions on GlobTel's motivations for entering into it. Tom Short, a GlobTel Manager who was a key actor in establishing the WS relationship, described the main motive to be a resource one. Scott, a Director of a GlobTel lab in the UK with which WS was closely aligned, also described the major reason in resource terms in order to develop 'headcount flexibility' in the UK. While WS saw the relationship as a means to 'go up the value chain', Barry, the MD of the UK lab, believed this expectation of WS was not realistic:

I don't agree with your analysis that when we started we looked at India in any other way than for cost reduction. We never saw it as a place for development of new technology. At that time [in the early 1990s], the strategy was to take the Digital Switching Product (DSP), a technology which was relatively well established in the UK, to Europe by taking advantage of the deregulation of markets over there, and to make DSP the preferred switch in Europe. The idea for us was to free up people and money over here given the constraints and conditions of the market by moving some development to Bangalore. So, the basic strategy was opportunistic.

WS was involved in switching and broadband projects, which L. Krishnan (the WS CEO) believed represented GlobTel's core activities, and into which they had entered by chance rather than design. Tom Short described the initial work in WS as being 'independent' without requiring too much everyday interaction between GlobTel and WS. Krishnan also described the initial projects to be small and stand-alone, without the support of independent communication links. Initially, a group of WS engineers went to GlobTel's North American office for six months and returned to Bangalore with memory reduction and regression testing projects that could be done by the Indian engineers logging into GlobTel's server in North America. With growth in technical expertise, WS started doing feature development projects, which Reddy, a senior WS manager, described as a movement from 'peripheral' to 'core' activities. Infrastructure upgradation facilitated this movement. Communication links were first provided in 1993. The WS Technical Manager Ram described this as an event of great significance because with the links in place the volume of work increased significantly and the number of WS developers increased from twenty-five in 1993 to 220 in 1996. In 1996, the GlobTel account was worth about US$8 million, WS' second-largest account and contributed about 16 per cent of their total software exports.

The relationship started to take on an exclusive status within Witech Infotech in a few years from its inception. The operation was housed in a separate lab (with restricted access) and employees wore GlobTel–WS logos on their badges. Mutual investments by GlobTel and WS into infrastructure reflected the growing intimacy in the relationship. While GlobTel made significant investments (to the tune of about US$ 1 million) in communication links and switching equipment for simulations, WS reciprocated by setting up an independent lab in 1995 and acquiring the required workstations. These mutual investments, despite the absence of a formal and legally binding long-term contract, were described by Krishnan as signifying a relationship in which 'the spirit goes much beyond the terms of the contract'.

WS' goal was to be designated as the 'preferred lab' where 'anything which they [GlobTel] want in India can be achieved through us' (Krishnan). WS did not see this relationship as a cost-cutting exercise because there was a continuous evolution in work, moving from testing to feature development. Ownership, a desired end-state, implied becoming fully responsible for the maintenance, enhancement and support of particular sub-systems transferred to WS. Under conditions of ownership, WS owned the architecture, the code and did all the approvals for any change in code without any backup from GlobTel. 'It is all with us' (Krishnan). However, ownership did not imply (as the term may suggest) control over IP, which was 100 per cent GlobTel owned, a clause built into the initial contract.

From the start in 1991, it seemed that GlobTel had consciously selected WS to be the 'hub' for their four Indian relationships. For example, GlobTel routed their telecom links through WS to the four partners, and WS conducted, on GlobTel's behalf, training programs for the other partners over videoconference to other GlobTel employees in North America and Asia. GlobTel located four of their expatriates on WS premises, three specifically to support WS activities and the fourth (Arvind, of Indian origin) to oversee GlobTel's India wide operations. Joe, one of the expatriates, described his mandate as to 'help them [WS] understand what GlobTel's expectations are'. As the unofficial hub, GlobTel consciously tried to integrate the WS group into their activities. Tom Short placed his group's annual operating plan on his web page so that could be accessed by WS in India and contributed to their sense of 'inclusion'. On their part, WS consciously attempted to replicate GlobTel's office environment and management style – for example, the manner in which they held project meetings. Krishnan remarked, 'Operationally we work as GlobTel and there is no difference. We have a similar set up to GlobTel.' To give GlobTel the necessary confidence in their long-term commitment, WS established a new telecommunications division in their organization structure and set up management levels that could be mapped to those of GlobTel. As Ravi, a project leader in WS said, 'The styles of management, the division of the hierarchy we are following is exactly like GlobTel. We have the same kind of structure.'

WS considered learning and adapting best practice models from others as central to its culture. It was therefore open to learning from GlobTel. Krishnan did not see copying as an issue and argued that 'the world over people compare, and there was no reason to be upset if they adapted best practices from GlobTel like their competence model'. He gave examples of how they had in the past adapted global best practices such as GE's Six Sigma methodology. However, in this process of adoption, some WS managers like Reddy felt that WS might be getting compromised as 'my people move more towards GlobTel than to WS and our identity is more with GlobTel'. She was concerned with the way in which 'the GlobTel identity has seeped into everything'. The identity was influenced through various mechanisms employed by GlobTel to disseminate information about themselves including General Information Sessions (GIS), newsletters, bulletins and access to their Intranet. As a result, people working on GlobTel projects 'were keyed into and better informed [and also curious] about events at GlobTel'.

Gradually, but decisively, through the efforts of both GlobTel and WS, a number of GlobTel's technical and management processes were introduced into WS. These helped to provide GlobTel managers with a sense of comfort that their processes and quality levels were being standardized in India, including systems for planning, reward and recognition and training. Some quotes describe the nature and mechanics of these management standardization efforts:

We plan out in detail, there are elaborate plans for each of these individual projects the way GlobTel plans. (Krishnan, the WS CEO)

I go to GlobTel once or twice a year, two–three weeks. We get exposed to and get to know their planning systems, and after working with it for a while, you get to know what to do and what not to do. It is the exposure and you get to know a lot of things. I think that is how I got to learn my management skills, looking at them, how they do, try to get their feedback, and then try to adopt it here if possible. (Reddy, a senior WS Technical Manager)

Actually, for appraisal we have a specific thing, what we say is project management, which includes client and project management. We have a similar appraisal program as they, and we fill [in] the same stuff. These standards are GlobTel predominantly. (Krishnan, the WS CEO)

Actually, we also have a spot award. It is famous, the GlobTel spot award, actually it is also their initiative. (Ravi, a WS Project Leader)

These processes of standardization were not without their own tensions. For example, around quality, Reddy lamented that 'despite being an ISO 9000 certified company, with excellent internal processes [we] cannot apply them since GlobTel does not accept them.' Reddy saw GlobTel as an extremely large and structured bureaucracy that was often overbearing on WS. She felt:

This bureaucracy often becomes a dictatorship. That is where the problem starts when they start saying: 'this is what I think you guys have to do. I don't want you guys to take up anything more. I want him to work two years on this job and I don't want him moved.'

While Reddy saw this tendency to dictate to be paradoxical when compared with their normally 'open' way of dealing with people, GlobTel felt this 'micro-management' was required since WS did not understand their expectations. As a result, as Al (an expatriate) remarked:

There was a need to get involved in the details of the way that individual deliverables are being run as well as the details of what we call people management, so whether this person is productive or not, whether that person should develop the feature, or [be] moved to another group.

Initially, the WS staff resisted the presence of expatriates, who themselves were also rather frustrated by the lack of progress. However, with time, the WS staff started seeing the value of expatriates in solving problems and became more accepting of them. The expatriates also became more realistic about what could be achieved. Fred, an expatriate, felt an initial source of conflict was the different standards in expectation levels. While an Indian designer would be happy to be given an independent phone, even if it might not work nine times out of ten, a North American would see an unreliable phone as similar to not having one. Fred found it frustrating to explain such contextual differences to his bosses in North America who could not understand, for example, why it takes so long to send a fax or for a decision to be taken by a WS manager. Gradually, a shared understanding was developed through the expatriates, and the frequent travels of WS managers to North America and vice versa. The resulting increase in the replication of GlobTel's practices and style in WS made some Indians slightly apprehensive. They felt this could erode their traditional strengths arising from the 'community' style of working as contrasted to the more individualistic North American style.

In trying to become GlobTel's 'preferred lab' and show commitment to telecommunications, WS consciously recruited people from the telecommunications industry. However, the issue of attrition continued to plague the relationship, as was the case in most firms in the Indian software industry. WS was experiencing an annual attrition of some 20–25 per cent. The temporal frames of reference used by GlobTel and WS to interpret attrition were different. While a person with 7 years' experience in WS was considered a 'relic', it was quite common to have 20-year 'veterans' in GlobTel. As a result, even though WS felt that it had effectively dealt with attrition to the greatest extent possible, GlobTel felt this was not enough. Over time, WS implemented a number of (mostly) GlobTel inspired initiatives to deal with attrition and developed a strategy of 'planned attrition' whereby additional resources were continuously created in anticipation of gaps caused by the departure of developers.

Larger industry dynamics also contributed to evolution in the nature and volume of work in WS. There was an increasing confidence worldwide about the level of expertise in Indian firms, which were now moving from the initial focus on maintenance as defined by customers to higher-value work where they had some degree of input in defining the project parameters. Another source of influence was the increasing introduction of packages in firms, for example Oracle and SAP. GlobTel, like many other firms, started to show a preference for standardized packages over doing in-house development, causing some uncertainty in WS about future opportunities. Within this context of growth and uncertainty, we describe the phase of growth in the relationship.

Growth and associated tensions (1997–1998)

After working together for 5–6 years, both WS and GlobTel felt that they had developed a level of mutual understanding and appreciation of each other's weaknesses and strengths. Increasing investments from both sides contributed to an evolution in work and also introduced various tensions. Reddy described GlobTel's increasing concern with attrition and low retention as unfair since GlobTel had many decades of experience in telecommunications, whereas WS was only a recent entrant. While she felt that GlobTel should clearly define their parameters and let WS take control of the process, GlobTel saw workforce instability impeding the upgrading of the relationship from a tactical to a strategic level. Venkat, the CEO of Witech Enterprise Solutions, saw GlobTel's concern as justified because of the situation of dependency of GlobTel on WS. The instability of the workforce exacerbated the problem.

WS was extremely sensitive to the need to develop a stable workforce, which was now about 250–280 strong. To retain them, WS developed initiatives such as the 'three-years' and 'five-years' policies whereby a WS employee on completing three years could spend a year in North America and after five years be given an option to become a GlobTel employee. In an interesting and radical move, GlobTel appointed Chandra, the Witech HR manager, on a two-year secondment as their India Prime reporting primarily to GlobTel. Chandra's mandate was to create common training programmes for all GlobTel's India partners and to standardize the skill sets of Indian managers to fit the mould of a 'uniform GlobTel manager'. Chandra expressed excitement:

I think it is a brilliant model been made by a top consultant and it specifies the level you are with respect to your competencies and what you are required to achieve. This way everyone speaks the same language. So, at the same level whether you are in Company A or B, there is the same level of skills required. There is standardization. By developing a 'Global manager', GlobTel can leverage it for different contexts because the cultural framework they are using is quite the same.

This move to secondment of Chandra was a crucial attempt by GlobTel to deal with HR issues like attrition within a common management framework. We discuss this further in the analysis section.

To cope with the pressure of developing new Internet-based technologies rapidly, GlobTel embarked on the strategy of acquiring start-up technology companies (for example, from Silicon Valley) rather than investing heavy amounts on in-house R&D. A proper response to these changes needed WS to attempt to enter new technology areas and develop patents, instead of passively waiting for GlobTel to transfer technology. Such a reorientation was difficult for reasons of proximity, as WS would need much greater exposure to the end-users primarily based in North America. In the light of worldwide uncertainty about the future of R&D in software development, Barry (the UK lab MD) felt WS' expectation of gaining IP ownership was unrealistic. It had been hard enough for GlobTel to give new technology work to their own UK lab, and so the possibility of Bangalore getting such projects was remote indeed. Venkat differed from Barry on these implications as he felt that although earlier projects involved proprietary knowledge about GlobTel's switching technology where control rested with GlobTel, the Indians had better possibilities with the open Internet technologies:

So on this telecom side this was really causing some kind of concern whether we will really be able to do it. But the good thing now is that telecom and datacom technologies are merging. Whereas telecom was more proprietary software, proprietary languages, etc., the datacom side is more of the Internet and more open languages or objects and similar technologies. It gives us an open field now to really come out with some kind of a chance.

The implications of GlobTel's 'right-angle turn' were beginning to show as the new budgets favoured new areas of wireless, and projects in switching were being cancelled. WS realized the pressing need to develop independent solutions, rather than waiting to be told what to do. GlobTel felt this could be difficult, as WS was not 'proactive' enough. In contrast, Reddy described 'proactive' as a much-abused term because it was not discussed in a context of a mutual flow of information for which both sides were responsible. WS thought it was difficult to be proactive since the intentions of GlobTel were unclear. GlobTel itself found it difficult for GlobTel to define long-term goals in the state of flux in the industry.

To deal with these changes and problems of attrition, GlobTel proposed setting up their own software development centre in India called GlobTel Software Overseas Development Centre (called GSODC). This move was met with obvious resistance from WS who saw this proposed centre as direct competition since it would mean that all 'crown jewellery' work would be done there. They would be given the legacy 'crumbs'. Venkat felt the GSODC proposal, if justified solely on the grounds of cost and attrition control, would not be effective because it would need to be managed by expatriates relocated to India at North American salaries, and problems of attrition and distance management were not going to disappear. He said: 'If you look at the whole Indian context, I don't think they can do better than we and hold back people from leaving.' Venkat also commented on the changing nature of work and the resulting pressure on the pricing basis of work:

My own feeling is that the current model is dying, maybe within two–five years. It is a very simple equation, if the cost goes up at the rate of 30 percent and the prices do not go up and it is only dependent on the rupee depreciation of 5–6 percent, no mathematics is required to find out whether in the third or fourth year you would be profitable or not. People are going to pay more dollars for higher value-added work, and how you are to provide that is the big question. We need a different model wherein we get value for money rather than based on counting the number of people.

On one hand, the existing pricing basis was being questioned as the legacy model was expected to die and a need was felt for models that could support 'small teams producing incredible value products' (Jim, a UK lab manager). In WS, even alternative business models were being suggested including risk and reward, turnkey and profit sharing. On the other hand, the situation in the industry and GlobTel made such changes difficult to bring about. For example, since the GlobTel lab in the UK, with whom WS was aligned, was itself facing uncertainty and downsizing, it was impossible for WS to expect a growth rate in excess of that of their 'parent'. The 'right-angle turn' route was placing additional pressure on GlobTel to change their technology strategy to acquisition rather than in-house development. Jim cautioned:

The right angle turn till now has been about acquisitions. GlobTel R&D is as much under threat as WS. James Brown [GlobTel CEO] has told the labs that if they do not come out with product announcements regularly, he will periodically come out with announcements about acquisitions. If they don't do that, I think their workload may move to zero by 2000. The same scenario is also applied to this lab, and so is not unique to WS. WS have to wake up to what everyone is facing.

It is within this context of uncertainty and change that the final stage is described.

Acceptance of the legacy route (1999–2000)

A key question in 1998 for GlobTel was whether WS would be used for legacy systems or for 'right-angle turn' development? Related issues were pricing, attrition and future expectations. GlobTel decided to use WS primarily for legacy system work rather than provide support for the 'right-angle turn'. GlobTel made a number of changes of personnel who were responsible for the India relationship, including Ravi the new head of the Indian operations (who took over from Ghosh). Reddy felt this move would give a stronger business focus than the earlier relationship-building efforts:

But with Ravi taking over, he has proved himself very well. He has a very good track record. He is coming here for a different purpose. He appears to be mainly business oriented where Ghosh [the previous in-charge] was primarily relationship oriented. He individually relates to a person, lots of feelings take place there. Ravi is very different. Different in the sense that nothing goes by perception or anything like that. I think it is a good change over from Ghosh to Ravi.

Through some unpredictable circumstances, WS benefited from GlobTel's 'right-angle turn'. Arising from a political process within GlobTel, projects from one of their labs in North America (with which another partner was aligned) were transferred to the UK lab with which WS was aligned. As a result, while in 1998 the other three partners experienced a decline in their business with GlobTel and a need to cut developers, WS increased their numbers of developers from about 275 to 350. The majority of the new projects for WS involved the legacy digital switching product (DSP) technology and outsourcing to them which would free the UK staff for new product development. Even though DSP represented legacy technology, WS saw this transfer as a solid future business opportunity since DSP already had a significant market presence. Maintaining it would provide steady 'bread and butter' work. Reddy rationalized:

I think the right-angle turn has helped us a lot. They [GlobTel] realized that they have to move on, they cannot hold on to a product hundred years old or whatever. The CO 24 [pseudonym for GlobTel switching products] is making US$ 500 million today. And SM 50 [another pseudonym] is also making a lot of money. So it is not going to go away. But they [GlobTel] may not stick to this because if they do their market opportunities in 10 years will go down. So they have to do more work in the packet and data area. So they need a place to give the existing products. It really is a good technology for us to understand; definitely it is not in the data area or other fancy areas. We realized that there is a business case.

A key challenge in accepting the legacy route was in trying to retain the young talented programmers who wanted to work on state-of-the-art technology rather than DSP. Another challenge was that DSP technology represented a 'shrinking pie' from which budgets would be constantly redirected to the 'right-angle turn'. So, although DSP legacy work might guarantee a certain level of work for future years, there was constant pressure that work would have to be done at reducing cost. Even though WS had accepted the legacy route, it was uncertain how long they would be able to sustain this work given the ongoing challenges.

In the next section, we analyse the process of evolution of the GSA relationship using the theoretical lens of 'standardization'.

4.3 Case analysis: a standardization perspective

Like many large MNCs in the 1990s, GlobTel was active in establishing offshore software development centres (like WS) globally to enable distributed software development. Coordinating work in these different centres raised the need for manifold and complex standards operating at different levels and locations. The extreme diversity and scope of standards that come into play in any GSA requires a conceptualization that is typically broader than past research that had focused on technical artefacts and infrastructures. For example, Hanseth (1996) describes standards with respect to basic communication protocols, their syntax, semantics and pragmatics of the information to be exchanged. Monteiro (1999) describes the implementation and deployment of relevant Internet standards, including the specification of communication protocols. Hanseth and Braa (2001) have examined in Norsk Hydro the implementation of the Hydro Bridge standard to improve coordination between various divisions and the corporate headquarters. Related research in the health domain (for example, Timmermans and Berg 1997) has focused on artefacts like medical protocols and the socio-political processes through which they are constructed and implemented. While such research has helped to understand how standards around artefacts or technologies are created, it does not explicitly account for the standardization of *management practices and processes* and how these are redefined through use. The analysis of this case focuses on two key questions:

- What are the nature and scope of standards that come into play in a GSA relationship?
- What are the mechanisms of translation through which these standards are created and integrated (or not) into everyday work – the process of standardization?

The analysis of these two questions helps to examine the broader question of the role of standards and processes of standardization in shaping the evolution of a GSA relationship.

The nature of standards

We have conceptualized standards as simplification and abstraction with the aim to define and communicate significant aspects of the processes, artefacts and structures across time and space. The aim of standards is to *universalize*. In GSAs, a wide range of standards comes into play covering at least five different domains: physical infrastructure; technical infrastructure; technical processes; education, training and technical support; and management processes. We provide some examples of the nature of standards in these domains.

Physical infrastructure

The physical infrastructure includes physical buildings, office layouts, coffee machines and even the badges worn by the WS staff working on GlobTel projects. WS established a separate building to house the staff and equipment attached to GlobTel projects and tried to create an office layout that replicated the set-up in Canada. These 'physical standards' helped first to create a unique sense of identity for the GlobTel relationship and secondly to provide the GlobTel staff when they were in India with a sense of comfort by being 'as if still in North America, within the GlobTel environment'.

Technical infrastructure

'Technical infrastructure' includes a number of items including workstations, replicated servers, networks, switches for testing, software tools for configuration management, programming languages, telephone lines, etc. For example, from the WS premises, a person could pick up the phone and dial a counterpart in Canada using just the extension number, as if in the same building. In just the same way a Canadian staff member could call WS. Although the replication of the technical infrastructure helped to improve the efficiencies of working – faster connectivity, for example – it also enabled a more seamless environment. GlobTel staff worldwide could operate within a common technological framework. GlobTel specified the use of a proprietary language as the development platform. Although a common language helped to standardize the technical implementation process, it made some developers feel they were being prevented from 'speaking' to others in the global marketplace, thus impeding their marketability and movement.

Technical processes

The technical processes include the software development methodologies, processes by which revised code was integrated into existing software archives and systems for quality assurance such as quality manuals and other documentation. There was a very detailed process on how a software developer after making revisions to a piece of code should send it to the central GlobTel archive, and the procedure by which that would be accepted, for example. Elaborate software development methodologies were in place that specified various 'gates' during the process. These gates signified when there would be particular 'hand-offs' of project deliverables, including software code, from developers in India to the central software archives. The methodologies adopted by GlobTel in most cases superseded those existing already in WS, typically developed in accordance to ISO and CMM quality specifications.

Technical and management knowledge

An important aspect of GSAs is the technical and management knowledge required to facilitate GSW. The technical domain includes knowledge of telecommunications, of specific switches manufactured by GlobTel, and of the proprietary programming language specified by GlobTel. Management knowledge includes understanding of North American and GlobTel culture, particular procedures and practices within GlobTel including systems of personnel appraisal, productivity measurement criteria for labs, and the matching of the organizational structure, management hierarchies and reporting relationships of GlobTel with WS. At a more meta level, as WS took on ownership responsibilities of certain products and features, they needed to develop a 'service mindset' and provide service at quality levels such that the customers could not differentiate whether WS or GlobTel were actually providing the service.

The above discussion emphasizes the different kinds of standards and the wide scope they cover – from the global–universal domain of software development methodologies to the very local level of employee badges. Some of the standards are open, for example, the CMM quality levels. Others are proprietary, like the platform on which GlobTel's systems were developed. Coordinating these very different types of standards is extremely difficult, involving various mechanisms of translation. Translation refers to the mechanisms used to 'interest [or impose on] others in your concerns' and describes the processes by which standards are introduced into the everyday practice of GSA. The introduction of these standards can be met with resistance, a reinterpretation of their meaning and an ongoing redefinition.

Mechanisms of translation

An analytical focus on standards and the various translations that surround them helps to emphasize how standards shift and the mechanisms through which these changes happen. As Hanseth, Monteiro and Hatling (1996) point out, a key challenge in standardization is managing the tension between the need for *stable standards* on one hand and for *flexibility* on the other. 'Flexibility' refers to the potential for further changes in redefining patterns of use and in the symbolic value the actors place on gaining a sense of control over the process. Our analysis aims at understanding these mechanisms of translations and how tensions arising from their implementation shape the course of the relationship. The following two main mechanisms of translation used by GlobTel were:

1 The establishment of WS as a 'hub' for GlobTel's India operations
2 The role of WS in the 'GlobTel manager' standardization effort.

WS as the 'India hub'

GlobTel were interested in standardizing their activities first across the four partners in India and secondly across the broader network of GlobTel's labs worldwide. From GlobTel's perspective, standardizing activities in India was desirable as it would help to make better use of resources, improve coordination and enable more effective sharing of information and knowledge across the partners. For a variety of reasons, including the sense of a loss of identity, WS did not always share the same perspective. The idea of a 'hub' is used as a metaphor to depict WS' role as the central coordinating point for GlobTel's activities transcending its four partners in India. GlobTel's standardization efforts can be understood at two levels: (1) internal to WS, and (2) across the four partners in India.

Within WS

Standardization was attempted through a variety of techniques – physical, technical and managerial. The independent lab served to create a unique identity of a 'GlobTel island', where even WS employees from outside the GlobTel group had restricted access. This unique identity was reinforced in many ways such as the WS staff wearing badges with 'GlobTel' logos and an office layout with the 'look and feel' of a GlobTel office in North America. On the surface, this segregation provided security, but at a deeper level it defined a predominantly GlobTel frame of reference, an identity, that was unique and visible in WS. This physical structure was reinforced by a number of other electronic mechanisms such as General Information Sessions, newsletters, bulletins and the Intranet. WS was inundated with information about GlobTel. The HR manager said she had introduced a system of performance appraisal orientation sessions once a month so that 'there is always a talk of it'. The pervasiveness and intensity of information made an Indian manager worry about her staff being more aligned to GlobTel than to WS, and being informed and interested in events and activities in GlobTel rather than in their home organization.

At the technical level, the information and communication technology infrastructure played a key role in developing standard frameworks of work in a material and symbolic sense. This reflects Akrich's (1992) argument that 'technical objects thus simultaneously embody and measure a set of relations between heterogeneous elements' (1992: 205). For example, the telephone link that enabled a WS staff member to ring a counterpart in North America and vice-versa as if they 'were within the building,' symbolically fostered for WS a sense of 'inclusion' in the relationship. The software development environment was established in WS such that work could go on in India 'as if it was taking place in North America'. GlobTel's quality processes and software development methodologies superseded WS' existing internal processes. Frequent videoconferences, emails and phone calls, reinforced GlobTel's way of working, such as the use of particular models for project monitoring. Frequent videoconferences enabled the increased 'presence' of GlobTel staff in India even though they were 'physically absent'. This electronic presence was further reinforced through the presence of North American expatriates in WS, more than in any of the other partners. These expatriates had the primary mandate of introducing WS to the 'GlobTel way of doing and thinking about things', and detailing how this could be achieved in practice.

A number of management systems were introduced in WS. These systems could be thought of as 'non-human actors' (Latour 1987) who spoke 'on behalf' of GlobTel in WS. These included GlobTel's systems of regular and continuous appraisal, spot awards for reward and recognition and honouring tenure. Systems of mentoring that involved attaching new recruits to the more experienced people in the project, used extensively in North America, were introduced in WS. Through a systematic and intensive introduction of these systems and information, the use of GlobTel terminology such as 'buddying', 'GIS', etc. became part of the everyday and also official vocabulary (for example, in appraisal forms) in WS. In many cases, systems first introduced into WS were classified as 'best practices' that were then spread to other groups within the company. This 'organizational' spreading helped to institutionalize and strengthen these practices.

External to WS

WS served as the key point of coordination of GlobTel's activities across the four partners. WS was the point through which communication links were routed. Initially, switches required for simulating the development environment were sent to WS, and subsequently to the other partners. From a management perspective, WS served as the entity through which GlobTel's management training programmes were conducted for the overall India operations.

As compared with the other Indian partners who resisted the idea of GlobTel expatriates sitting in their offices, WS had expatriates located almost permanently on the premises, including a Director responsible for the overall India operations. The use of expatriates and consultants was a particularly strong technique to develop stronger 'inscriptions' of GlobTel standards. We use the term 'inscriptions' in the same sense as Callon (1991) when he says that 'an inscription is the translation of one's interest into material forms' (1991: 143). Interestingly, many of the expatriates were of Indian origin with many years' prior experience with GlobTel in North America. They were thus expected to have a sufficient understanding of the local culture to enable effective communication with the Indians, and to be integrated into the broader GlobTel system. They could thus understand the GlobTel perspective and would usually concede the superiority of Western management practices and standards. These expatriates defined their mandate in India to 'introduce the GlobTel way of working in WS', to make 'WS understand better the nature of GlobTel's expectations', and to make the Indian system more 'objective' and 'accountable' like their own. Edstrom and Galbraith (1977) identified the significance of sending managers from head offices to their subsidiaries, coupled with a systematic pattern of socialization as a means of control (Kamoche 2000). WS expected the expatriates to serve multiple roles in exercising control through micro-management and also to be a useful resource for solving mostly technical problems. The expatriates, coupled with many experts from North America, served as crucial *agents of change* to introduce and reinforce the standard GlobTel culture in WS, and through WS to the Indian partners.

Training sessions were important translation mechanisms since they built similar approaches for developing technical competence, and created common cultural and linguistic frameworks. There is an integrative relation between management training and the development of a shared culture. Training serves as a tool for the transmission of culture, which in turn furnishes the rationale for training (Kamoche 2000). Organizational values incorporated in training courses assume legitimacy by becoming part of the knowledge required for job performance and career advancement. In international management, training efforts need to go beyond simple skill and competence formation to the more complex domain of how knowledge can be transmitted and interpreted across cultures. The complexity arises from the fact that such knowledge has both functional and symbolic values since it appeals both to the managers' self-interest (career advancement) and to their sense of ideology, in this case inscribed in Western management practices.

WS and the 'GlobTel manager'

A key initiative taken by GlobTel to integrate culture and various HR initiatives was to create a 'standard template' for management through development of a 'Global manager'. This reflected GlobTel's attempt to create a universal template of management that could be used as a basis to define, guide and manage the work in the four Indian organizations and compare this work across the broader global network. This template could also serve as a basis to coordinate the diverse activities of software development projects that require simultaneous interactions with many different groups of people and organizations. GlobTel selected WS to help implement this standard template by taking on secondment the WS Human Resources Manager (Chandra) who continued to sit in the WS office. Chandra described her task in the following terms:

There is standardization. GlobTel has so many D-Level managers, for example, all over the world, and using these standardized systems they can take a quick look at the level of the skills set to determine the overall competency. While the roles and competencies are the same, the managers are of course different. So, India and UK can be merged and made one. I believe the standards and competencies of GlobTel managers are higher than what [they are] here. We are trying to culturally change some of the behaviours of the managers here. Then at the same level, depending on the behaviours that are exhibited, GlobTel can possibly reward them. By developing a 'Global manager', GlobTel can leverage it for different contexts because the cultural framework they are using is the same. Culturally, we can change them and make them more aware. The model is generic in nature. You may have the most beautiful eyes, nose, and mouth, but put together the impact may not be so good. It depends on how things fit together. The model is generic and it covers 22 spheres.

Chandra tried to introduce within WS and with the other Indian partners various 'best practices', including a quantified system for performance measurement called the 'Performance Dimensions Dictionary' (PDD). The PDD served as a new 'common language' to describe employee competencies and serve as a reference document to develop standard techniques for identifying and measuring them. It also specified how to use these measurements as a basis for performance appraisal. 'Competencies' referred to a set of measurable performance criteria designed to enable understanding 'on-the-job' behaviours of professionals and to help them 'improve' and thereby enhance their chances of success in the company. An internal document described the role of the PDD as follows:

It is intended to be a reference document for people in GlobTel globally who want to identify ways to improve performance and to establish objective assessment of performance. The common reference point will help provide the objectivity that is needed to ensure fairness, and connect people processes such as recruitment, development, appraisal and training right across the organization.

A number of other quantified systems were introduced to develop measures of productivity, competency measurement, performance appraisal and for attrition monitoring. Models of quantification potentially serve to develop powerful inscriptions because of the inherent belief that managers have in numbers, and the standard basis it provides to compare processes across time and space. An interesting example of this quantification is the PDD used as a means to quantify systems of performance appraisal and to enable its uniform spread by its being placed on the Intranet. The PDD consisted of nine parameters represented in a matrix form with different weightings depending on the level of the person. The supervisor computed an aggregated index as an overall performance indicator. Employee competencies were graded on a scale of 1 (lowest) to 4 (highest). With this quantification and standardization, it was expected that managers from one organization could be compared with others and their own performances monitored over time. This system of quantification had implications for larger managerial processes such as recruiting, dealing with attrition and computing the average competency level in the organization.

A number of other systems of quantification were also introduced. GlobTel introduced a system of monthly reporting for the Indian partners that gave the productivity of the lab in quantitative terms. After some resistance, this system of productivity reporting was formally introduced. A local practice of not having such written reports was thus subsumed within the standard global framework of lab productivity reports. With these reports, GlobTel managers in North America could potentially interrogate the Indian labs about their monthly performance and compare them with other labs worldwide.

Another quantitative system was the self-assessed 'user satisfaction reports'. Initially, this system was also resisted by some of the Indian labs but subsequently accepted on GlobTel's insistence.

A larger aim of these various translation mechanisms was to create a technical, cultural, managerial and physical framework in which GlobTel employees in both North America and India could feel the same as in North America. Such a structure was designed to create more efficient mechanisms of coordination and to bring economies of scale through the sharing of resources. From the WS perspective, these standards were supposed to upgrade management practices by introducing systems that would help them engage with telecommunications and GSW more effectively in the future. WS was to feel like a 'virtual GlobTel' lab and part of a larger worldwide lab network. However, attempts to standardize came with their own tensions, which shaped the process of the relationship.

4.4 Standardization, tensions and the GSA process

GlobTel's strategy of making WS their India 'hub' provided the broad framework on which the relationship was built. By establishing an independent telecommunications division within their organizational structure, WS sought to create a common basis to standardize practices that could be compatible with GlobTel and to ease the processes by which the relationship could proceed. The early establishment of a GlobTel lab in WS helped to provide exclusivity and a sense of identity to the staff. The early standardization of the physical and technical infrastructure helped GlobTel build a sense of confidence in WS, which contributed to a rapid development in the volume and level of work. Of the Indian partners, WS was one of the first to be given projects leading to ownership. Ownership came with its own tensions, however, in particular related to workforce stability and new technologies.

WS seemed to be willing to serve as the Indian 'hub', and be guided by the GlobTel way of doing things. WS' resistance to GlobTel's attempts to micro-manage seemed relatively milder and less explicit than that of the other partners. On the contrary, it seemed to express a sense of openness and eagerness to learn and adapt GlobTel 'best practices' and methodologies. This adoption took place at multiple levels from organization structure to project management practices, to employee reward and recognition schemes down to even the employee badges and the common language used around the building. These processes reached a kind of 'peak' when WS seconded their HR manager for two years to GlobTel where she would report on WS (and India activities) to GlobTel. This represented an extremely powerful way for GlobTel to develop stronger inscriptions of their standards in WS and India in general. The dominant GlobTel system of information, activities and structures, however, made some of the WS managers worry about the erosion of the WS identity among the workforce, especially the younger staff.

A key tension at WS concerned the validity of the pricing model. It had been considered appropriate in the initial stages when work was routine and WS was learning the ropes of GSA and telecommunications. But it became questionable when work involved ownership transfer. WS started to question payment based on time and material, and to suggest it should be based on the value and knowledge they brought into the process. This was not acceptable to GlobTel for various reasons, including IP concerns, expected increase in costs and fear of a loss of control over the process.

Two external but interconnected events significantly influenced the course of the relationship. The first was the Internet, that changed the technological focus in the industry and put a question mark on the future of DSP technology and its supporting budgets. The second was the political restructuring that took place in GlobTel's corporate headquarters. GlobTel had various labs in North America (Montreal, Ottawa, North Carolina, etc.) that were each aligned with different labs in the UK, France, Israel and India. The various North American labs typically dealt with different technologies and products and so realignment in technological focus affected the various counterpart labs and, through the network, the offshore activities. Through a series of events, a corporate decision was taken to transfer a significant amount of the DSP technology from a lab in Montreal to the UK lab aligned with them. And since WS was aligned with the UK lab, a corporate decision was taken to 'transfer this transferred technology' to WS. WS gained as a result, and their business showed a growth as compared to the other Indian partners who showed a decline in 1998. This growth, however, created another tension – that of WS accepting work in a low-growth area (of DSP) when the rest of the industry was showing a rapid growth (50–60 per cent per year). This caused a further tension, that of trying to retain the best talent without being able to offer them work on what they perceived as the latest technology in the industry.

It is interesting to examine the nature of the standardization efforts and how they progressed over time. In general, an increase in standardization initiatives was accompanied by increasing levels of work (from initial bug-fixing to feature development to ownership), both in quality (from peripheral to core technologies) and in volume (large-scale ownership transfer). The broad trend seemed to be that with a steady increase in the content and quality of work, there was an increasing need to standardize first the technical domain and progressively the management processes. Although the initial focus was on technical and infrastructure standards, with time the focus shifted to processes and practices, especially to systems of HR management. A more advanced and extreme form of standardization came with the attempts to create the 'global GlobTel manager' framework.

Expatriate managers drawn primarily from GlobTel's International R&D Group (GRDG) were key actors in guiding the various standardization initiatives, including infrastructure, office space and management and technical know-how. With the processes of ownership transfer gaining ground, WS started having more direct linkages with the UK lab it was aligned with, and viewed GRDG as redundant. GRDG had succeeded in establishing a robust infrastructure that so strongly inscribed the GlobTel way of working that its own role was seen to be redundant. The GRDG set-up was slowly phased out in India; people either resigned or returned to the corporate fold of Glob-Tel. Somewhat paradoxically with increased standardization GRDG, the agency that had been responsible for setting up this infrastructure, was withdrawn. Metaphorically, GRDG served the role of setting up the scaffolding of the building, and once that was done, the scaffolding became invisible.

The metaphor of scaffolding provides an insight into Latour's (1999) question about 'what is gained, what is lost, and what remains invariant in the process of translation?' Latour raises this question in his discussion of the notion of a 'circulating reference', and the manner in which the idea of standardization is tied up with the concept of 'invariant':

A reference is not simply the act of pointing or a way of keeping. Rather it is our way of keeping something constant through a series of transformations. What a beautiful move, apparently sacrificing resemblance at each stage only to settle again on the same meaning, which remains intact through sets of transformations. The rupture at each stage of the 'thing' part and its 'sign' part. The details are often lost, and what remains is the horizon, the tendency. Reduction, compression, marking, continuity, reversibility, standardization, compatibility with text and numbers – all these count infinitely more than adequatio [does this mean resemblance?] alone. No step – except one – resembles the one that precedes it, yet in the end when I read the field report, I am indeed holding in my hands the forest of Boa Vista. (1999: 56)

A key point that Latour seems to make is that standardization involves a process of small translations where some form of a 'global standard' is introduced at the local level and activities are subjected to a redefinition with reference to this standard. This introduction and comparison takes place through a series of translations involving a process of dialectical interaction between the local and the global, where something new is gained, something is lost and something remains the same. The invariant part of this process of translation reflects the strength of the standard. In Latour's Boa Vista case, the Munsell number serves as a reference that is quickly understandable and reproduced by all the colourists in the world on the condition that they use the same compilation. It permits a crossing of the threshold between the local and the global. In our case, what remains invariant can be conceptualized in terms of the scaffolding used to set up the structure of the relationship including support of a number of different technical, management and physical routines.

At one level nothing has remained the same, as happens with everything with the passage of time. Initially, WS did not have a strong expertise in telecommunications or in GSA, and the relationship with GlobTel helped them to develop it. Through a series of contested and uncontested translations, WS incorporated a number of GlobTel's management processes to create a 'WS–GlobTel hybrid' reflecting a culture of GSW and telecommunications expertise, albeit with a dominant GlobTel reference. What is gained is this new technological and business expertise, a steady business of DSP legacy projects and management values shaped strongly by GlobTel. What may have been lost in this process of translation are some high-quality staff who did not want to be limited by GlobTel's legacy work, an erosion or redefinition of the WS identity and perhaps some of their work practices being superseded by GlobTel's processes.

While gaining and losing are hazards that organizations have to engage with in the present context of globalization, a complex and ongoing question is what remains invariant in this process of translation. When the scaffolding is removed, the building is left behind. Even though people who live in the building may change over time, and it might be used for purposes different from those previously intended, the structure has a defining influence in the nature of these changes and redefinitions. We can see GlobTel's contribution to the creation of such an infrastructure within WS, including the physical building, the technological infrastructure, the expertise, the management structures and processes and a certain culture of doing telecommunications work within a largely North American approach. Although the specific people in the building move away, the structure has an influence on where they go, what kind of new people come in and the kind of work that goes on.

There is thus both an ongoing discontinuity and continuity, with a stronger tendency towards continuity. This tendency is something that people knowledgeable about the software industry in India would associate with WS – a reputation for being relatively passive, being conservative with an extremely good business sense, and keeping a low profile despite extraordinary financial successes. The company has been a key influence in shaping the larger trajectory of the industry. With the decision to accept the legacy route, WS took the path to enter areas that made excellent business sense rather than jumping into risky ventures that would involve cutting-edge technologies. From that perspective, accepting the legacy route can be seen to reflect and reinforce some of the existing 'low-profile' tendencies, some of which, it could be argued, have remained relatively 'invariant'.

The case points out the extreme complexity inherent in the infrastructure that supports GSW. This complexity emphasizes Hanseth's (2001) point that new understandings and strategies are required for complex arrangements such as information infrastructures and not seen as information systems that tend to be stand-alone rather than part of complex networks. Standards and processes of standardization, as we use the terms, are enmeshed in these complex physical, technical and managerial infrastructures. They reflect Fujimura's (1992) notion of 'standardized packages' that relate to technologies adopted by multiple social worlds that construct new and temporally stable definitions of these technologies. This notion of standardized packages is broader than Star and Griesmer's (1989) notion of 'boundary objects' that enable coordination of work by members of different social worlds having different perspectives and agendas. Standardized packages reflect a broader conceptual and technical workspace that combines several boundary objects, and serves as an *interface between various social worlds to facilitate the flow of resources*. This space that defines a GSA concerns the physical spaces of the two partners, the electronic spaces created by technologies like videoconferencing and the Internet, and the various standards and processes that link the work practices within these physical and electronic spaces and with other nodes of the network.

The case points to the futility of attempting to build 'universal standards', since they are constantly redefined, negotiated, reinterpreted and applied differently. Hanseth and Braa (2001) have argued that attempts to create universal standards often lead to the opposite effects of creating complex and 'non-standardized' systems:

Each time one has defined a standard which is believed to be complete and coherent, during implementation one discovers that there are elements lacking or incompletely specified while others have to be changed to make the standard work, which makes various implementations different and incompatible – just like arbitrary non-standard solutions. This fact is due to essential aspects of standardization. The universal aspect disappears during implementation, just as the rainbow moves away from us as we try to catch it. (2001: 264)

Knowledge or technology is shaped and also shaping the meanings and actions of the heterogeneous network of actors surrounding the design and use of standards. The question is not how the technical content of particular standards is best applied in universal settings, but how different local particularities interplay with these standards to redefine their meanings at both the universal and local levels, how global standards are embedded or not in local practices and how actors respond to these improvisations. As Timmermans and Berg (1997) argue, although universals exist they only do so as *local universals*, embedded in local infrastructures and practices, paradoxically as a multiplicity of universalities. It is not a matter of dismissing universal standards or celebrating local particularities, but of developing a pragmatic balance (Rolland and Monteiro 2002) that blends the universal and local in particular contexts.

This case analysis provides ideas on how such a balance can be aimed for by understanding the mutual linkages between standards and their role in shaping the process of the GSA relationship. These interlinkages are complex since the growth of technical standards helps to upgrade the level of work and to take the relationship to a stage of growth requiring a more sophisticated standard such as that of the 'global manager'. External events, like industry developments, contribute to further destabilization of existing standards and create the need for new ones. These dynamics emphasize the quest for universal standards to be akin to 'catching the rainbow' (Hanseth and Braa 2001). Understanding them can help develop standards that can sensitively support the relationship, and yet be flexible enough to respond to local needs. A first step to this is the need to reconceptualize standards:

1 A primary aim of standards is to support the *universalization of activities*, and with it to facilitate some form of 'mass production'
2 Standards should be seen as a means of *simplification and abstraction* that defines and enables communication across actors separated by time, space and cultures
3 Standards are formal–informal, explicit–tacit, external–internal, and emerge incrementally through a series of *politically negotiated translations*
4 Standards apply not only to the technical, but also extend to the *management and physical domains*; standards across these domains are interconnected, and there may exist a hierarchy of standards with physical and technical standards serving as a precondition for the setting up of management standards

5 There is an ongoing interplay between the need to apply more *'global' standards* and the need to provide *flexibility at the local level*

6 The process of defining and enforcing standards takes place through a series of *'translations'* and in each such move some aspect of the standard changes and other aspects remains; what remains 'invariant' is of interest and is what we conceptually identify as the standard

7 Translation processes are the key to shaping the process by which a GSA relationship evolves over time, which in turn define the nature of further standardization efforts; understanding this *complex interplay over time* is crucial.

BIBLIOGRAPHY

Akrich, M. (1992). The description of technical objects, in W. E. Bijker and J. Law (eds.), *Shaping Technology/Building Society*, Cambridge, MA: MIT Press

Bowker, G. and Starr, S. L. (1999). *Sorting Things Out. Classification and its Consequences*, Cambridge, MA: MIT Press

Callon, M. (1991). Techno-economic networks and irreversibility, in J. Law (ed.), *Sociology of Monsters: Essays on Power, Technology and Domination*, London: Routledge

Edstrom, A. and Galbraith, J. R. (1977). Transfer of managers as a coordination and control strategy in multinational organization, *Administrative Science Quarterly*, 22, 248–63

Fujimura, J. H. (1992). Crafting science: standardized packages, boundary objects, and translations, in A. Pickering (ed.), *Science as Practice and Culture*, Chicago: University of Chicago Press

Hanseth, O. (1996). *Information Technology as Infrastructure*, PhD thesis, Gøteborg University
    (2001). *Information Infrastructure Development: Installed Base Cultivation*, Working Paper, Department of Informatics, University of Oslo

Hanseth, O. and Braa, K. (1998). Technology as traitor. SAP infrastructures in global organizations, in R. M. Hirscheim, M. Newman and J. DeGross (eds.), *Proceedings from 19th Annual International Conference on Information Systems (ICIS)*, Helsinki
    (2001). Hunting for the treasure at the end of the rainbow: standardizing corporate IT infrastructure, *Computer Supported Collaborative Work*, 10, 3–4, 261–92

Hanseth, O., Monteiro, E. and Hatling, M. (1996). Developing information infrastructure: the tension between standardization and flexibility, *Science, Technology and Human Values*, 11, 4, 407–26

Kamoche, K. (2000). Developing managers: the functional, the symbolic, the sacred and the profane, *Organization Studies*, 21, 4, 747–74

Knor-Cetina, K. (1999). *Epistemic Cultures: How the Sciences Make Knowledge*, Cambridge, MA: Harvard University Press

Latour, B. (1987). *Science in Action*, Cambridge, MA: Harvard University Press
    (1999). *Pandora's Hope: Essays on the Reality of Science Studies*, Cambridge, MA: Harvard University Press

Leidner, H. (1991). *Fast Food: Fast Talk*, Berkeley: University of California Press

Monteiro, E. (1999). Scaling information infrastructure: the case of the next generation IP in the internet, *The Information Society (Special Issue on the History of the Internet)*, 14, 3, 229–45

O'Donnel, S. M. (1994). *Programming for the World: A Guide to Internationalization*, Englewood Cliffs, NJ: Prentice-Hall

Rolland, K. H. and Monteiro, E. (2002). Balancing the local and the global in infrastructural information systems, *The Information Society*, 18, 2, 87–100

Star, S. L. and Griesemer, J. R. (1989). Institutional ecology, 'translations', and boundary objects: amateurs and professionals in Berkeley's Museum of Vertebrate Zoology, 1907–39, *Social Studies of Science*, 19, 387–420

Taylor, D. (1992). *Global Software: Developing Applications for the International Market*, New York: Springer Verlag

Timmermans, S. and Berg. M. (1997). Standardization in action: achieving local universality through medical protocols, *Social Studies of Science*, 27, 273–305

5 | Global software work: an identity perspective

5.1 Introduction

The significance of identity in GSAs

In the ongoing processes of globalization, transformations in individual and group identities are a subject of much contemporary debate. Giddens (1991) writes that while globalization can be understood at an institutional level, changes that occur as a result of it can directly impact at the individual level. He writes: 'transformations in self-identity and globalisation are the two poles of the dialectic of the local and the global' (1991: 32). Thus, a distinctive feature of contemporary life is this increasing interconnection between the two extremes of globalizing influences on the one hand and personal dispositions on the other. Evolution of organizational identity goes hand in hand with transformations of individual identities, which has a strong association with the 'rootedness' or a sense of 'place' that individuals experience (Godkin 1980). As will be discussed in chapter 6 on the dialectics of space and place, ongoing social transformations impact the sense of place with manifold influences on individual identity. Castells (1997) also emphasizes the dialectical relation between the net that metaphorically represents a universal instrumentalism based on the network logic of society and the self that is rooted in historic, particularistic identities that primarily are socially and geographically place-dependent. Individuals respond to the relentless pressure of feeling uprooted in the globalizing world through what Castells describes as the 'power of identity'. Castells (2001) argues most emphatically that identity is central to the processes of social change in contemporary society:

Social action in our time largely depends on the construction of identity. It lies in the crises of political institutions such as the nation state, and of the institutions of civil society largely linked to the nation state such as political parties, labour unions, professional associations – the whole world of corporatism. (2001: 10)

While social and political implications of globalization and identity have been subject to numerous debates, its impact on individual and organizational identities, set in global business environments, has not been adequately explored in IS research. Walsham (1998) has argued that micro-level studies of issues like identity can help to develop deeper understandings of the relationship between IT and social transformation. GSAs provide an interesting context to study aspects of identities, since they involve new forms of organizational boundaries as compared to traditional firms where strict demarcations existed between internal and external relations. For example, as described in the GlobTel–Witech case (in chapter 4), the Witech group functions as if it were a 'virtual extension' of GlobTel, where the developers work with the GlobTel infrastructure, tools and project management practices, and consciously try to forge a 'GlobTel-like' identity. Linked with this aim of creating a GlobTel identity, internally there is the effort to configure the Witech group like an 'island' and isolate it from other groups (for example, those dealing with American Express and General Electric). This isolation is enforced through a number of mechanisms such as having groups dealing with different clients located in separate buildings (in the same campus), having restricted access to other buildings, wearing client-specific badges (e.g. of GlobTel) and even contractually preventing staff from changing groups. The ongoing management effort is to remove the *external boundary* between GlobTel and Witech and, paradoxically at the same time, establish new *internal boundaries* to distinguish different groups. As firms globalize, and new clients and geographical markets are addressed, the different boundaries are further conflated, and actors continuously need to shift between multiple and continuously evolving work, technological and social contexts. This shift places pressure on individuals, and changes in identity are intricately intertwined with these movements. In this chapter, we argue that these transformations in identity significantly influence the evolution of the GSA.

The ongoing interaction between internal and external boundaries in GSAs makes it necessary for actors to reflexively monitor the separations and develop a coherent sense of identification. In the software domain in a globalized setting, where professional attachments are strong and individualized, internal and external boundaries are continuously negotiated. Creative knowledge workers prefer to work in organizations where they believe there is freedom. The suppression of freedom leads to resistance, often expressed by workers leaving the organization. It is no surprise, therefore, that the Indian software industry has an annual attrition rate of about 25–30 per cent. Some Indian software firms try to prevent attrition by not allowing Internet access to their young recruits because the company fears they will use the Net to find employment in North America. Such prevention only frustrates individuals who thrive on and celebrate openness and creativity. The ongoing and reflexive relationship between openness and secrecy that typically exists in organizations is magnified in the case of GSAs. On the one hand, GSAs can be seen to dissolve boundaries by providing various choices of movement, while on the other, they create dilemmas of entering into unknown domains and settings, which reinforces the need for boundaries. To operate and move effectively between these multiple spaces, actors continuously seek to reconfigure their identities and reflexively monitor the consequences of their actions. Identity, while at one level in a constant state of flux, at another level remains extremely complex to reconfigure because it is historically and socially situated.

Figure 5.1 Organization culture–identity–image linkage

Conceptualizing identity

Conceptually, we analyse identity in relation to its linkage with organizational culture and image. Hatch (1997) argues that culture, identity and image form three inter-related parts of a system of meaning, sense making and action that defines an organization both internally and externally:

> Organization culture needs to be considered in the development and maintenance of organizational identity. How we define and experience ourselves ... is influenced by our activities and beliefs that are justified and grounded by cultural assumptions and values. What we care about and do defines us to ourselves and thereby forges our identity in the image of our culture. The symbols of our culture become important sources of meaning and identification. (1997: 358)

We argue that the dynamics of this relationship between organizational culture, identity and image is intricately intertwined with the process of evolution of a GSA. For analytical purposes, we schematically depict the culture–identity–image relationship in figure 5.1 (adapted from Hatch 1997) and subsequently elaborate upon its intertwining with the GSA process. In figure 5.1, a three-way system of meaning of culture–identity–image linkage is depicted, and discussed further below.

Organizational culture

Organizational culture has been an extensive topic of discussion within IS research (Robey and Azevedo 1994). A functionalist approach has been dominant in the study of culture with authors discussing national, organizational or work cultures in a manner that treats them as different objects that can be separated. Within these categorizations, attempts have been made to further isolate functional areas like control structures, communication practices, or dimensions (such as masculinity–femininity) and subject them to a cultural analysis. Such analysis has led to various prescriptions on how culture can be 'functionally managed' (Peters and Waterman 1982) (see also our discussion of culture in chapters 8 and 9). In attempting to counter such functionalist accounts, there have been a number of efforts to develop anthropologically inspired, situated and interpretivist understandings of culture (Avison and Myers 1995). Such approaches view culture as a process that is socially constructed and extensively negotiated and achieved (Westrup *et al.* 1995).

An interpretivist perspective, in contrast to a functionalist one, does not see organization culture as a variable that can be controlled, but instead as process that is constituted by and at the same time helps to constitute a *socially constructed context*. A functionalist approach treats culture as something that an *organization has* – for example, the work of Hofstede (1980) who equates nations with different cultural traits. The alternative interpretive approach treats culture more in the terms of what an *organization is* (Smircich 1983), reflecting a relatively more 'spiritual' kind of phenomenon. Interpretivists argue against Hofstede's 'scientific' view and suggest that there is no necessary and inevitable alignment between culture and the nation state (Myers and Tan 2002).

The 'has' and 'is' distinction of culture each represents a duality that is incomplete. The 'has approach' de-emphasizes the nature of managerial agency, while the 'is' focus does not give adequate importance to the structural context within which culture is constructed. Giddens' attempts to dissolve such dualisms in the study of social life have been articulated in structuration theory (1984), and have in recent years been applied quite extensively to the study of information systems. We draw upon some of Giddens' ideas in an attempt to develop a perspective to unify the 'has' and 'is' approaches to the study of culture. Actors, as simultaneous members of different work- and non-work-related social systems, draw upon various rules and resources they interpret to exist in these systems in the process of articulating agency. For example, as members of globalized and high-tech software firms, developers interpret creativity as a key resource of their membership. They draw upon these resources in the process of creating new rules or reinforcing existing ones (such as informal dressing and working late at night). By situating individuals within different social systems of which they are members, we acknowledge the *structural constraints* that come as a result. This reflects the 'has' aspect of culture conceptualized in the form of rules and resources that individual actors interpret as associated with these systems. By focusing on agency and its interpretive basis, we emphasize the 'is' basis of culture, and the power that agency has to redefine some of the structural constraints. The recursive structurational relationship that Giddens describes to link agency and structure provides us with the conceptual basis to develop a perspective on culture.

Culture, viewed in structurational terms, sensitizes us to the various social systems of which actors are members as well as the rules and resources that are drawn upon in the process of articulating agency. For our analysis, we are interested in agency as related to the construction of *organizational identity and image*. 'Image' is concerned with the expression of the organization for the external constituency, while 'organizational identity' refers to the processes through which members develop identification. These processes of agency and image construction take place in terms of a subtle instantiation of rules linkage. The source of new rules or changes in rules is the outcome of patterns of behaviour repeated over a period of time. It is through this conceptual link between structure and agency as described by Giddens that we analyse the interconnections between culture, identity and image.

Organizational identity

Organizational identity is the second node of the three-way system of meaning and action. In contemporary times, the organization has become an important source of identification and has taken up, in large measure, the roles of religious and caste groupings of traditional societies (Burke 1973; Christensen 1995; Whetton and Godfrey 1998). Identity is thus an important concept in contemporary organizations; it has been broadly described to be what members perceive, think and feel about their organization, representing what is central and enduring about an organization's character (Albert and Whetten 1985; Scott and Lane 2000). These authors conceptualize stability as an 'essential' quality that provides guidance for responses in a turbulent internal and external environment. Such a view implies identity to be stable and conditioned within the broader and enduring organizational environment. This emphasis on stability has in more recent times been criticized (Gioia, Schultz and Corley 2000) and is in sharp contrast to Castells' (1997) discussion on the transformative aspect and potential of identity. Castells argues that through the expansion of capitalism in current times, the roles of traditional institutions are being dissolved. Organizations are being forced to reconfigure themselves, and individuals 'must rebuild meaning from the inside out, not from their minds but their practice' (Castells 2001: 10). We take Castells' perspective of identity not as something that is enduring and stable, but as something which is being *continuously redefined* through the action of individuals in situated circumstances.

This linkage between identity and agency is in line with Fiol's (1992) emphasis on the relationship between cultural rules, identity and observable behaviour and language, speech, acts and words: 'People's organizational identity provides the context within which behaviours are linked to the rules that give them meaning' (1992: 200). Identity serves as a focus of efforts to gain competitive strengths by providing organizational members with a stable sense of themselves while operating in an extremely turbulent environment. When effective, identity so constructed helps to provide a new context for linking behaviour to rules and a basis for redefinition of strategies. Identity in new and growing organizations like ComSoft is thus consciously formulated, shaped by the actor's own interpretation of what is appropriate identity. This action at the level of discursive consciousness selects particular resources and rules in terms of aspects of cultures that are perceived to be helpful in the development of an appropriate organizational identity. Faced with the diversity of situations and challenges, creating structures incorporating and reinforcing chosen elements of contextual resources becomes an essential responsibility of senior management.

Organizational image

Organizational image has been defined as the manner in which organizational members believe others see their firm (Dutton and Dukerich 1991). Organizational image thus involves externally produced meaning about the organization, which is important in shaping internal organizational processes. For example, Dutton and Dukerich (1991) describe how the New York Port Authority was forced to take action on the homelessness problem as a result of community pressures expressed through a negative organizational image. The organization's interpretation of their own external image triggered action, and was filtered through organizational identity. Image is thus different from identity in its external orientation and concerns the feelings and beliefs about the company that exist in the mind of *external audiences*. Image is communicated externally through mission statements, logos and press announcements. This action is shaped by the need of organizations to adapt to different business, cultural and technological contexts. Senior managers play a key role in conceptualizing and articulating an organizational identity and ensuring a strong identification both internally within the organization (identity) and externally to the world at large (image). Changes in organizational image help to reconfigure internal processes of identity formation. Organizational culture provides the actors with a sense of identification internally, and recognizability and marketable value externally, which reflects the image of the organization (Dutton, Dukerich and Harquial 1994). Hatch (1997) describes organizational identity and image as a self-reflective product of the dynamic processes of organizational culture:

Culturally embedded organizational identity provides the symbolic, material form from which organizational images are constituted, and which can be projected outwards. Organizational images are projected outwards and absorbed back into the cultural system of meaning taken as cultural artifacts and used symbolically to infer identity. (1997: 360)

In summary, our conceptual framework is built on a three-way system of meaning that comprises organizational culture, identity and image. Culture provides the context that shapes and is shaped by the articulation of agency through a subtle instantiation of rules linkage. We are interested in two forms of agency. The first is an *internally directed* agency that concerns the construction of identity and linked to the experiences of individual actors drawing from their work- and non-work-related memberships. The second concerns an *externally directed* agency that reflects the organization's position *vis-à-vis* the stakeholders at large. Both identity and image are conceptualized as being mediated through organization culture. The culture–image–identity system of meaning helps to develop insights into the multi-faceted, dynamic nature of a GSA and the processes through which it evolves.

In GSAs, where the individuals are in constant interaction with the external world of clients, industry and the marketplace, the boundary between image and identity tends to be blurred. As in the case described in chapter 4, the image of Witech as a 'virtual GlobTel lab' is not only an expression of its image in the global community, but also becomes a source of meaning and identification within Witech. This image–identity linkage is mediated through a culture that encourages and actively adopts 'global best practices' into its own organizational processes that are reflected in the quest of replicating this 'virtual GlobTel lab'. The image of a stable organization is fundamentally dependent on the identity of organizational members, who find meaning and relevance in working in the organization. Their sense of meaning and identification is in turn strengthened by a stable image. Through reflexive processes that link identity, culture and image, the cultural context both constrains and enables agency directed towards the construction and articulation of image and identity. This link is reflected in Covalski *et al.*'s (1998) description of the manner in which managers engaged in training and mentoring draw upon the discourse of the 'professional autonomy' image of the 'Big Six' public accountancy firms to shape the identities of organizational participants. Actors resist these attempts, a situation which leads to a partial redefinition of identity and image.

In this chapter, we are interested in examining the process of evolution of a GSA relationship and the role of identity in shaping this process. Identity is conceptualized within a broader framework of meaning that links it to culture and image. Processes of transformation of GSAs, we argue, are fundamentally linked both to the sense of organizational identity experienced internally by actors and to the perception of image experienced by the external world. Two sets of processes become important. The first concerns the *mediation of culture* in the construction and expression of both image and identity. The second concerns how the *external constituency experiences this image* and the processes through which the construction of identity is further shaped. The objective of the chapter is twofold. First, it is to describe the processes of transformation of the GlobTel–ComSoft GSA. Secondly, it is to examine how the evolution of the GSA is intricately tied up with expressions of identity and their transformations over time.

5.2 Case narrative

Communication Software Ltd (ComSoft for short, a pseudonym), a relatively small Indian software firm, focuses on high technology, including engineering software and telecommunications. Their strategy is different from that of the traditional Indian software firms that have typically focused on business software services. The ComSoft business divisions include: wireless, switching, networking and R&D. ComSoft reflects an unusual mix of high-technology capability coupled with self-confidence and deep pride in its national and cultural roots. This combination is unusual since the culture of firms in developing countries is generally associated with difficulties in adopting new technologies. The case narrative is presented in two broad phases of *inception* and *transition*. 'Inception' relates to ComSoft's initial phase of start-up and growth that took place under the mentoring of GlobTel. 'Transition' relates to the phase in which ComSoft tried to break free from the GlobTel umbilical cord and to grow, drawing upon a redefined sense of their identity and image.

Inception (1991–1996)

ComSoft's inception in India was largely supported by the link it had with GlobTel and the mentoring and support provided by Paul and Ghosh described in chapter 3. These mentors saw in ComSoft a prototype of the vision they had for the Indian technology sector – creative, professional, high-tech and Indian. As a policy, GlobTel did not engage in 'body-shopping'-type projects and encouraged only those that could be sustained over time in India itself. This policy was greatly appreciated by Raj Moitra, the ComSoft founder and CEO, who was striving to emphasize the 'Indianness' of his high-tech and creative company. Like GlobTel's other Indian partners ComSoft, too, in the initial stages of the relationship, was trying to understand GlobTel's technology. At first, ComSoft was engaged in projects that were relatively routine software bug-fixing and feature-development-type activities on older telephone exchanges. Slowly, with time, ComSoft was assigned tasks in the more challenging areas of wireless communication. These tasks provided a better scope for the technological innovation that they were striving for. To enhance their capabilities in new technology domains, Moitra approached two faculty members, Dr Sandip and Dr Vikas, in the computer science department of the Indian Institute of Technology (IIT), a globally recognized leader in technology education. Both Sandip and Vikas were acclaimed in their subject area, and had decided to stay in the country rather than migrate to prestigious universities in the West. They appeared to be driven by an ideology similar to ComSoft's, that of wanting to contribute to technology development from 'within' rather than 'out of' India. These common interests helped to seed an interesting and growing relationship between ComSoft and IIT that was described by the HR manager of ComSoft, Ms Deepa:

We have a very close relationship with IIT because when the company started out, the founders had gone to IIT and had requested a few people for consultancy assignments, and two of the professors – Dr Vikas and Dr Sandip – helped us. We had a link with IIT because they were partly working with us and continuing to work till they completely moved out to our organization. Despite our being a low-profile company in the market, we are able to attract high-calibre professionals from the universities. We have eleven PhDs with us.

ComSoft, after its initial confidence-building tasks with GlobTel, was bolstered by the involvement of the two IIT faculties, first as consultants and then as full-time employees. The company became increasingly proactive by developing ideas and designing solutions, trying to elevate the GSA from a vendor-contract mode to a 'true partnership'. A senior VP said:

We have been working with GlobTel in that relationship essentially because of the nature of the job. But now we start becoming proactive and go with our own ideas and design entire sub-systems. Then the kind of relationship will be different. We could still be a vendor in some sense, but then we will be adding more value and earning more money. And the business model will not necessarily be one of just getting cost, but it will also mean getting royalty. It will not be per man-days but in terms of the business.

Key to this process of evolution of the relationship to a partnership mode was ComSoft's vision of *technological excellence* and *self-reliance*. The top management in ComSoft consciously tried to develop an organizational identity centred on 'technological excellence'.

The close association with IIT and the involvement of its professors provided the platform on which such a high-tech image could be built. ComSoft channelled a significant
15–17 per cent of their funds into research and promoted an environment where the
engineers were encouraged to compare their work with the best in the world. ComSoft's
vision of technological excellence reflected the ethos of Silicon Valley start-ups, including
a flat, matrix kind of organization structure where creativity was valued and junior programmers were encouraged to voice their opinions in team meetings. ComSoft adopted
a similar open-door policy favouring merit, in sharp contrast to typical family-owned
businesses in India where relationships to the owning family are significant. Moitra
described ComSoft's 'flat' structure:

Treating people in the right manner costs money. If I travel, I want our people to stay the same way
as I do. Those things cost money. So there is a pressure and simultaneously there is pressure on the
competition front. There are a lot of companies coming up, and still cost is, I believe, driving a lot of
the corporations to set up shop in India. So, the challenge is that we see that we move out, we move
up the value chain and truly become partners rather than just sub-contractors.

The idea of technological excellence was closely and intimately tied up with the notion
of self-reliance, and ComSoft proactively tried to shape an identity of self-reliance and
'made in India'. They developed a mission statement of *unleashing Indian creativity*
in the global arena, reflecting the attempt to appeal to the sense of passion of young
Indians who desired to see their country as a leading site for cutting-edge technologies.
The 'made in India' was a desired state of identity that the ComSoft management
deliberately tried to cultivate. As Ms Deepa, the HR manager, described it:

Our company has many people who are very passionate about India, extremely passionate. And they
have a larger purpose, a vision of making India a stronger place, making India a centre of excellence
in the world. This is a very, very intense feeling that people have over here and they feel that if we can
change lifestyles over here then people will not run to the US. So can we become a model of excellence,
can we become so profitable that we can change lifestyles over here so that [with] the same things that
US can provide, excellent work and excellent money, we can keep people here. And that is what we
can do given the larger purpose of making India a centre for excellence for technology.

The company consciously tried to appeal to the 'Indianness' of the organization by
taking pains to emphasize the 'Indian way' of celebrating festivals like Diwali involving
traditional attire and food. The company also strongly emphasizes the importance of
family and family relationships to create a sense of togetherness. This is in contrast to
the Sierra case in chapter 7, where the Indians' family orientation was seen to impede
professional commitments. The values of Indianness, high technology and self-reliance
were consciously cultivated to help articulate the vision of unleashing Indian creativity. The CEO and the senior executives like Dr Vikas and Dr Sandip were proactively
involved through workshops and seminars in creating an organizational buy-in for the
vision statement. Interestingly, the consultant who proposed this vision was an Indian
living half the year in Silicon Valley and the other half in India. This was reflective of the

hybridized nature of the vision combining Silicon Valley creativity and drive with Indian ethos and values. ComSoft hoped to achieve this vision by building independent products and technologies that in turn would take the GSA from the current vendor-contract situation to the status of a true partnership. Moitra described his firm's ambitions:

We have to catapult ourselves into providing complete solutions that GlobTel's customers are looking for. I am talking to their direct Project Line Managers to support the customization of their products and make it a success in India and in this region. I am just not restricted to India alone. In this region we can play a major role in supporting and customizing the products. So, I believe those two are the major challenges that we are working with. And educating our people, especially the young ones who are joining just out of college, that learning about a product is equally important as trying to crack an algorithm or something like that.

The conscious cultivation of an Indian high-tech organization served another instrumental role of contributing to control attrition. First, the firm tried to recruit those developers who indicated a strong identification and desire to work primarily from India. This identification was further reinforced through various mechanisms that emphasized the value and enthusiasm of 'being in India'. A project manager reflected on why the attrition rate in ComSoft was lower than the industry average:

People do come and go, but I will say that it is very low-key in ComSoft. ComSoft is stable. The industry standard (of attrition) is 25 percent, right. I don't see that happening here. This is because it is a small company, so basically it is the small company culture that attracts people. You are close knit and you have fewer problems, and salaries are good, and the work that they are doing is pretty much on the frontier level, all these things are there. Most of the people who are leaving are not leaving to join other companies, but are leaving to go abroad. The high-tech profile is one of the things that attract a lot of people.

In summary, the inception of ComSoft was greatly supported by GlobTel, with the conscious aim within both ComSoft and GlobTel that the identity would be that of a high-tech, creative, university research kind of set-up. This identity was pursued strongly by the CEO Moitra, together with some high-profile faculty members from IIT, first as consultants and then as full-time employees. As confidence and expertise in ComSoft grew, they began to strain at the umbilical cord which some felt was tying them to GlobTel. Feeling the strain in many areas of the relationship – ranging from the kind of projects they were being given to the budget constraints – ComSoft consciously started to develop strategies to make a transition to a new independent identity.

Transition (1997–2000)

We use the term 'transition' to describe events and processes that relate to ComSoft's attempts to break free of GlobTel's 'umbilical cord' and grow independently in different markets and technological domains.

The goals of technological excellence and self-reliance consciously inculcated by ComSoft helped it to develop a sense of purpose and identity. However, in making a transition to a new identity, the limits of its relationship with GlobTel gradually became apparent. GlobTel was reluctant to adopt the new technologies and patents developed by ComSoft; GlobTel preferred the earlier system of paying ComSoft on a 'time and material' basis. Further, changes in the industry because of the Internet were forcing GlobTel to redefine their technology focus and to make a 'right-angle turn' from a 'voice' company to a 'data and voice over data' company. ComSoft, despite the interest of a few sponsors in its abilities with new technologies, found the typical attitude in North America towards changing from payment based on 'time and material' to one based on royalty or patents to be negative in general and lukewarm at best. A senior VP described their efforts to change the pricing basis:

> We have to renegotiate relationships where necessary. We have already started this process with GlobTel informally, but it will be formalized soon. Till now it has not been an issue because we were not generating intellectual property during the course of the projects. But now we will, and therefore, we have to negotiate the rights.

There were changes in GlobTel around this time of transition catalysed by the 'right-angle turn'. As discussed in chapters 3 and 4, the 'right-angle turn' referred to GlobTel's attempts to reconfigure itself from a 'voice company' that primarily manufactured telecommunication equipment to a 'data and voice over data' company that emphasized communications, both data and voice. The uncertainty and turbulence that existed in the industry meant that managers, even the new set of senior managers of Indian origin, were incapable of enlisting and supporting Indian capabilities in their worldwide business, especially new technology development. The Indian software industry had received increasing worldwide attention, and there were several new clients who contributed to a substantial rise in earnings for Indian companies. GlobTel, at one time, paid one of the highest rates for software work in India. These rates had, however, not kept pace with the times and over seven or eight years GlobTel became one of the less attractive paymasters in the industry. GlobTel's new strategy was to acquire small companies, a faster way to obtain new technology than to develop its own R&D. New GlobTel managers were driven by opportunities for marketing their telecom equipment in India: it was felt that this opportunity had not been adequately exploited in the past. A senior ComSoft manager described this changing focus within GlobTel:

> They have a new head here – Ravi. So any change that is not formal, not valued on either side, is pretty opaque. We have some changes in the key people. Paul leaving has had its impact and with Ravi taking over, obviously it will create upheaval. Ravi has got a different charter, a different interest. He is more interested in developing the Indian market; he has got responsibility for both development and marketing in India. John is well plugged into action in GlobTel. His interest is in developing the Indian market. He is not worried about other departments (concerned with software development). He is looking at development partners from the point of view of what value they add to the goal of doing business in India (and not only do software work).

Frustrated largely by the GlobTel situation within this phase of dynamic change, ComSoft strengthened their efforts to become independent of the redefined business and technological focus of GlobTel. This process was reinforced by the 'Japanese factor' that helped ComSoft further to explore and also reinforce their identity of self-reliance. They displayed a sense of clarity and purpose, as reflected in this comment of a senior ComSoft VP:

We are not so bothered because for us in the long run this model [time and material] itself is dying. For us, we know our goal is to have a risk and profit sharing model, and by the end of this year we will get out of the time and material mode. We want to change this model. The issue of rate changes to the new model is secondary.

Interestingly, in pursuing this new model, ComSoft were not limiting their growth and expansion strategies to GlobTel and North America. In their search for markets for ideas and products, ComSoft discovered a different set of attitudes in Japan, which they felt was more compatible with their philosophy of technology development. With the initial experimentation in Japan being a success, ComSoft expected an exponential growth in their Japanese business in the future. A senior ComSoft VP speculated:

Right now GlobTel is about 30-odd percent of our business. In the future, the percentage might remain the same. It might even drop because other customers will be added. Japanese customers have just started picking up. We are at the threshold where it is just taking off and then it will take off like crazy. [The Japanese] take a long time to start working, but once they have decided, then they pull all stops. It has happened in one company already.

The potential in Japan made the ComSoft managers positive about the future. They believed success in Japan could lead to changed relationships in North America, including an increased market base with a wider set of clients. Another executive speculated on the nature of these changing relationships:

We want to look to other customers in North America. So far we do not have other customers in North America because you have to go and prove yourself. Unlike Japan, in America there is still the mindset that we want to work with Indian companies to cut cost. That is the first reaction mindset at least. Deeper it could be different. But the customers need more convincing. So, now that we have reasonable success with a new kind of arrangement in Japan, we are more confident of convincing people in North America. And once we launch these products, automatically our credibility will increase. And for services also, we will get more customers in North America.

These changing relationships were expected to lead to ComSoft developing a greater technological competence in telecommunications, which could contribute to significantly redefining the nature of the ComSoft–GlobTel relationship. Another ComSoft manager discussed this possibility:

There are a lot of similarities, but there are differences also. Because by the time we started working with the other companies, our technological maturity was much higher than what it was when we started working with GlobTel. And also we had already started taking in a lot of technological initiatives and so on. When we started with GlobTel we really didn't know any telecom at all. Now that is the reason why we are trying to change this relationship with GlobTel, because now it is possible.

Nearly three years after the initial experimentations with Japan, ComSoft has realized that it has not been able to make inroads into the Japanese market as rapidly as it had initially expected (and hoped). Japan comes with its unique and complex set of problems with respect to operating software-outsourcing relationships, some of which we discuss in chapter 9. Also at the level of some programmers and developers in India, there appears to be less enthusiasm about software skills in Japan and the USA is seen as the major source of learning for software technologies. ComSoft managers also spoke in complimentary terms about Israel, admired for its technological skills and approach to US markets.

ComSoft's Japan office manager, Japanese by origin, had strong opinions about ComSoft's strategy in Japan. He felt it should be different from the North American approach. He was of the view that 'unleashing Indian creativity', while successful in the USA, needed to be redefined for the Japanese market, as the managers there preferred technology that had been approved in the USA. This feedback to ComSoft India came through different channels, including our research reports. ComSoft consciously took the decision not to advocate the mission statement so expressly and instrumentally and instead keep it more subtle and in the background. In the meanwhile, the organization had also grown tremendously and they moved from offices in a residential neighbourhood into a 'university-like' campus structure that could house 3,000 employees. Also, shaping the changing context was GlobTel's 'right-angle turn' strategy. ComSoft, together with three other partners, were given the maintenance of selected legacy systems. Although such work was against the general ethos and values of ComSoft, they accepted it; they rationalized that the sheer volume of this legacy work would provide them with a 'bread and butter' operation while allowing them to continue exploring Japan and also new markets like China and Israel.

The telecommunications expertise developed through the GlobTel relationship gave ComSoft the confidence to explore these new markets in Japan. With growth over time, the relatively small and high-tech culture of ComSoft was also redefined with an organizational structure based on divisions to help coordinate their global operations. ComSoft management took the success on the NASDAQ of Infosys, a large Indian firm, as a model to emulate for the future. Following this model implied that ComSoft would have to focus more on the commercial aspect of the business as contrasted with the earlier focus, at times even obsessive, on high technology. In the meanwhile, inspired by the rapid growth of the telecommunications marketplace, ComSoft took the brave and futuristic strategic path of tapering off their existing areas of work (in Electronic Data Automation) and focusing exclusively on telecommunications. This redefinition of strategy was reflected in the change in the company name to 'ComSen' and the redefinition of the mission statement from 'unleashing Indian creativity' to 'making connections'. This transformation also reflects Castells' (2001) argument that identity is created and redefined by what people do in situated circumstances and globalization is enacted within an instrumental information network.

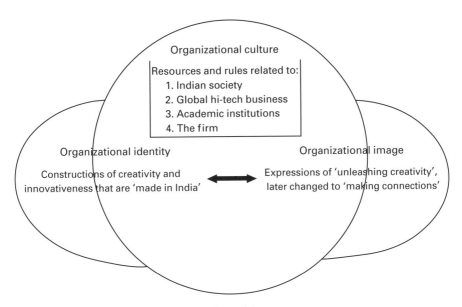

Figure 5.2 ComSoft–GlobTel GSA: culture–identity–image linkage

5.3 Case analysis: an identity perspective

The cultural complex

In this section, we present a case analysis that is based on the conceptual framework outlined in figure 5.1. The application of this framework is schematically outlined in figure 5.2 and then discussed. To build and apply an analytical framework that links identity, culture and image, we conceptualize managers as members of a cultural complex that comprises multiple social systems, including *Indian society, global high-tech business, academic institution* and *the firm*. We summarize in table 5.1 the rules and resources that can be seen to constitute these systems and their role in shaping agency. We then elaborate on the relationship between identity and the evolution of the GSA.

Indian society

Traditionally, Indian society has been stratified on functional lines, a system that has been reinforced by government policy of reserving job and university admissions for people belonging to 'backward castes'. This stratification is supported by the joint or extended family that networks people who share similarities in terms of caste, language and kinship. The basic resource stems from paternal authority and the rules of paternalism or an authoritative superiority tends to govern the system. Family priorities are seen to be important considerations in the shaping of choices such as location of employment. Staying with parents in India as contrasted to going abroad and leaving the family is

Table 5.1 Social structure and managerial agency in the GSA relationship

Social system	Indian society	High-tech business	Academic institutions	The firm
Dominant structure	Family and community	IT businesses and professional groups in a globalized context	University and general academic environment	ComSoft and the work and non-work environment
Basic resources	Importance of family as cornerstone of Indian society and paternal structure Importance of education in family Hindu worldview and its unifying systems of belief	The 'Silicon Valley' culture of global high-tech business Role of GlobTel, a global giant Globalization with American and other developed countries open increasingly to outsourcing work The 'Silicon Valley' culture hybridized in the Indian ethos Indian software industry dynamics with successes creating visibility and attracting quality manpower	University system, with premier status accorded to intellectual values Emphasis on self-reliance; it was with this aim that IITs were created by the government	Organizational structure and hierarchy Company policies Physical building in which it is housed Various organizational and project-related processes
Basic rules	Family priorities influence career and location choices Education, especially technical, as key to family status and financial rewards Devotion to knowledge and commitment Integration of work and family lives	Focus on high technology, innovation and risk taking Anti-hierarchical organizational structure to reinforce high-tech image Indian software industry that favours growth and is dominantly export oriented, seeking to establish a significant global presence	Nationalist principles based on the theme of developing technological self-reliance Professional codes, especially related to computer science Ambivalence towards technology as a generator of economic returns; interested in technology for its own sake and not for its end-use	Emphasis on creativity Principles of participation Equality of status Recruitment processes that Emphasize high-tech high-quality rather than volume Dress codes, emphasize informality

a constant conflict faced by software developers working in GSAs. Indians are often described to view work as a form of duty to their families (Sinha and Sinha 1990), which is reflected in the preference for personalized and family oriented relationships. In government and other traditional business organizations in India, as described by Sahay and Walsham (1997), family and caste systems have a significant influence in work areas. This is generally considered a sign of lack of professional development because often it is not merit that is valued but family membership.

Private businesses in India have often been described as drawing upon family values such as the respect for paternal authority in the conduct of business, with less emphasis on developing 'professional' management practices (Khare 1999). In contrast to this negative perception of family, ComSoft explicitly extends the family metaphor to the work arena. Relationships between members of ComSoft have a strong sense of family inscribed in the work practices: celebrating in the company environment festivals that are normally celebrated at home, for example. This invocation of family and community values helps to provide a sense of comfort to 'techies', and it is directed outwards with the aim of limiting attrition. Actions of senior managers such as staying in similar hotel accommodation as junior developers, participating in Indian cultural festivals, arranging workshops that encourage participation of junior staff, and being part of the prevailing discourse of high-technology research all help to create a cultural context that values and reflects creativity with an 'Indian slant'. Developers who would typically want to migrate to the USA are made to feel that they are able to do high-technology work in a 'USA-like' environment but importantly in the comfort of their own homeland. By being able to do similar work, actors help to create a cultural context that has dramatically different properties as contrasted to traditional Indian firms. This cultural context *mediates the construction of identity and also the linkage with image.*

High-technology business

Some of the resources of the high-tech global system come in the form of the Silicon Valley values that are embraced by Indian software firms such as ComSoft. These cultural values are manifested in rules that encourage risk taking and innovation and that value creativity. Sahay and Walsham (1997) use a similar structurational framework to analyse the implementation of geographical IS in the Indian government. The global high-tech business was not considered important in shaping the cultural context of the government managers, which was more significantly influenced by their 'national' membership. Resources arose from the socialist approach of the government and were manifested in the rules and formal procedures of the bureaucracy. The different sources of rules and resources emphasize the need for a situated approach to study circuits of reproduction, since in the same country (India) managers from different organizational affiliations (public and private sector) base their identification on varying motivations and sources.

The ComSoft case is played out in an environment of a dynamic, high-technology global business that mirrors and also draws upon the technology-based high-tech cultural values associated with Silicon Valley (Saxenian 1996). Actors draw upon a globalized mix of values that reflect a hybrid of the Silicon Valley kind of quest for technological excellence coupled with the Indian ethos. These resources are thus not just Indian, but have elements of North American high technology (as seen in GlobTel) and academic values intricately intermingled. Technology plays a key role in shaping these cultural values as ComSoft developers seek development opportunities in high-tech areas on a par with their US counterparts. The invocation of high-technology values helps to establish a 'comfort level' for the staff: Indian cultural values (for example, related to family) are refashioned or reinterpreted in Silicon Valley terms. Identity definition is constituted within a context shaped by global software opportunities and the Indian software talent pool.

The global software opportunities for ComSoft came in the form of the MNC giant GlobTel which aimed to draw upon the resources of Indian software professionals in order to meet their own shortfall. GlobTel's efforts in India were spearheaded by expatriate Indians who subscribed to the sense of Indian competence in software development and championed the idea of developing technology 'in India' rather than 'out of India'. But since these values needed to operate within a global setting to assure quality and acceptance within the broader GlobTel set-up, Paul and Ghosh encouraged ComSoft to construct an organizational culture with elements of Indian and Silicon Valley values. This could help to provide confidence to the senior GlobTel managers and potentially address issues of staff attrition. ComSoft managers appreciated Ghosh and Paul's mentoring efforts and this relationship became an important conduit for the construction of both identity and image. On one hand, Ghosh and Paul were seen as 'insiders' who provided ComSoft with inputs on how identity should be fashioned. ComSoft's implicit trust in Ghosh and Paul meant that their suggestions were given serious consideration in shaping organizational policies. On the other hand, Ghosh and Paul also served as 'outsiders' based in GlobTel and North America who could tune in and understand how the external constituency interpreted ComSoft's image and provide important feedback for change in the articulation of identity and, with it, in the presentation of image.

Academic institutions

Authors in the past have said that an Indian view of technology needs to be understood in context of an intellectual system that gives primacy to knowledge development, often resulting in the technology being valued for its own sake with little consideration for its ends (Saha 1992). The IITs from which ComSoft draws a significant component of its manpower are a product of the post-independent Indian nationalist outlook that prominently espoused self-reliance as the founding principle for growth. ComSoft draws upon these values to articulate a vision of 'unleashing Indian creativity'. However, within the global nature of ComSoft's business, this theme had to be interpreted in the appropriate context. This provides an interesting hybrid of innovation, technological excellence and self-reliance that is valued by individual developers. ComSoft started from lower-end bug-fixing work and moved progressively towards independent solutions in the telecommunications domain. In the course of setting up management systems to enable this progression of work, ComSoft decided to recruit people with a strong technology focused academic background and drew support from senior GlobTel managers who had confidence in Indian academic and technological strengths. The particular GlobTel Director (Ghosh) assigned to nurture ComSoft had a vision of developing ComSoft in a 'university-like' R&D structure with a strong focus on creativity and learning. A combination of these influences contributed to ComSoft's clear articulation of organizational identity in terms of a striving to unleash India's creativity in the global arena.

With growth and an identity that valued creativity, ComSoft soon realized the technological and financial limits of its GlobTel GSA. Under pressure from the global financial slowdown and the churn in the technological context GlobTel, caught up in its own turmoil of change, was unable to provide ComSoft with opportunities for creative work. Seeing a limited future with GlobTel, the ComSoft management decided to scale down GlobTel's work and redirect its efforts to the apparently more promising avenues provided by Japanese companies. The nature of creativity, as the Japanese factor shows, cannot be standardized to the North American frame of reference but has to be reinterpreted and rearticulated in different ways for varying markets. Through their initial explorations in Japan, ComSoft soon realized that the Japanese preferred technology that had been first accepted and tested in North America. This realization contributed to ComSoft consigning the image of 'unleashing Indian creativity' to the background, making it less explicit, and finally changing it as a part of fashioning a 'new look' that emphasized 'making connections'.

The firm

The structure of the firm represents ComSoft and its work and non-work environment. It is the environment in which individual actors find primary membership in their work settings. The ComSoft environment is constructed through a number of different elements that have a bearing on the individual actors and how their agency is shaped. The organizational structure and the reporting relationships within the organization favoured participation and informality. Senior management consciously encouraged junior developers to participate in meetings and say what they believed important. Formal company policies did not differentiate between the senior and junior staff, as they would stay in similarly priced hotel rooms (often shared), fly economy class and have similar expenses rates. This flat organizational structure was modelled on a Silicon Valley start-up firm, hybridized with the metaphor of family values. Commitment was developed and expressed not so much because of the formal contractual agreements that employees had with the firm but from the excitement and energy of a young group of people working together in a family environment. The physical structure in which they worked (a residential house) helped to reinforce these family values modelled in a broader framework of a global high-tech software firm.

ComSoft grew rapidly; the number of developers increased from less than 100 to nearly 1,000 in 6–8 years. There was subsequently an expansion in the geographical spread of operations and a need not only to change the physical surrounding of the residential environment but also to reorganize the structure and practices. The firm then moved to a large corporate office in another part of the city in 2001. It adopted a divisionalized structure in which the different business divisions reported in to the corporate headquarters in Bangalore. In the new facility, ComSoft has tried to maintain its characteristic relaxed and informal culture by having one large canteen area in which the senior and junior staff could eat together. However, how the changes in the physical setting and the organizational structure and practices will shape individual agency is an open empirical question for the future.

Culture–identity–image linkage

Rules and resources, or rather their interpretation and appropriation by human actors, provide the basis to analyse the linkage between human actions and the cultural context within which they are situated. Organizational identity and image are both developed in a self-referential pattern which influences, on the one hand, the choice of elements of the cultural context considered appropriate, and on the other, an image that is expressed to the external constituency. This image is an object of conscious articulation of managerial agency that shapes organizational identity internally among members. Through these self-reflexive processes, both organizational image and identity are shaped by the cultural context that is instantiated in action and drawn upon the rules and resources that actors see as important. Such a conceptualization is compatible with Boland's (1996) argument that an institutional-level analysis should not be made on the basis of shared belief systems and motivations, but in terms of 'circuits of reproduction' that represent *situated practices of actors* that feed back upon patterns at the institutional level: 'The forms or patterns of an institutional level analysis are not reducible to the meanings supposedly shared by individuals' (1996: 696). Identity is involved in these self-referential relationships between agency and cultural context as agents implicate it in the interpretation of rules. Identity in turn influences the processes through which the resources and rules are implicated in managerial action. Organizational identity can be seen to influence what aspects of the environmental stimuli are and are not noticed, and how particular organizational agenda should be shaped such as the kind of image that should be presented (Stimpert, Gustafson and Saranson 1998).

An understanding of situated practices and how they develop over time provides the basis to understand the culture–identity–image linkage and to infer the manner in which identity is intertwined with the evolution of a GSA relationship. Identity has been envisaged not as a shared belief held in common by organizational members, but instead, in terms of managers' actions in situated practices that tend to reproduce patterns or forms at the institutional level. The identity development process involves both intentional activity and unintended consequences of action. To create a cultural context in which desired patterns of behaviour may occur, senior ComSoft managers explicitly and implicitly formulated rules of behaviour relating to travel and meetings and organizational policies relating to personnel appraisal, etc. These rules were expressed consciously and unconsciously both internally to organizational members and externally to other constituencies through different mechanisms like mission statements, corporate logos, text carried in the websites and various material artefacts. In the early days, when ComSoft were in the process of building up a culture of an organization that emphasized family values, we have seen that they consciously took the decision to run their office from a large building in a popular residential neighbourhood so as to provide a 'home-like' environment to their employees. The cultural values symbolized by this building and its location provided an important source of identification for ComSoft staff.

The cultural context being shaped by the ComSoft managers, with active support from Ghosh and Paul, drew upon resources from four key areas:

- *Indian society* helped them to emphasize the family and parental structure. The resources drawn from Indian society were refashioned to reflect the global high-tech image that ComSoft wanted both to articulate consciously to the external constituency, and to strengthen their identity.
- The *global high-tech* structure helped to emphasize the 'high-tech' aspect that was developed and projected through Silicon Valley values. ComSoft subscribed to and reinforced this ethos by the presence of PhDs from the prestigious IIT that had a global standing as a high-quality technology institute.
- The role of *academic institutions* with a high technical status was the third important resource for providing identification to members and helping to shape the 'unleashing Indian creativity' image.
- The firm provided the fourth key aspect of the cultural context, reflected in the organizational structure and policies and the formal and informal practices acted out in the physical workplace.

In a new and growing organization like ComSoft, the role of senior managers is important in establishing circuits of reproduction of desired patterns of action by organizing social occasions like Indian festivals and workshops to discuss Indian values and philosophy alongside talks about high-technology areas of wireless and embedded design, for example. Through the instantiation of resources and rules as policies and practices, a common stock of knowledge is created for members to draw upon in the course of their actions. In ComSoft, this stock of knowledge was geared towards creating a cultural context that supported members wanting to work in India and motivated to be innovative, implying an identity with its own significance and distinctions.

Actions that create a sense of organizational identity and those that enact a desired strategy are inseparable. ComSoft, which started out as a GlobTel protégé executing assigned tasks, progressively discovered and constructed its identity in a global milieu. From the point of view of GlobTel managers, this can be seen as an unintended consequence since ComSoft asserted its independence, which then influenced the GlobTel managers' own relationships. The early image of 'unleashing Indian creativity', although apparently an effective mechanism to retain talented developers, had different implications in the global milieu. Japanese managers perceived this image as being largely negative since they preferred technology that had been approved and tested in the North American marketplace rather than in India. This feedback from the external world had rapid implications on organizational identity. ComSoft managers consciously placed the 'Indian creativity' aspect in the background and emphasized 'making connections'. This shift was also part of a larger process of organizational transformation of ComSoft in which they refocused their area of business operations primarily on telecommunications. It also imparted a greater commercial viability to Comsoft operations, making them more attractive to shareholders. These transformations needed to be fashioned within a different cultural context and a redefined emphasis on image and identity. ComSoft had also grown rapidly and, symbolic of their redefined focus and image, we have seen that it moved out from the earlier smaller building in the residential area into a large corporate office.

Understanding the link between image, culture and identity emphasizes the point that organizational identity cannot be viewed in static terms as it is *embedded and constantly re-expressed in action.* At one level, business organizations can be viewed as systems with relatively sharply defined boundaries arising from physical location, industry membership and markets serviced. The sense of '*us* versus *them*' is an essential ingredient of the organizational identity and image linkage that shapes competitive functioning and at the same time provides members with a sense of belonging. However, at another level, businesses have to operate in global environments, seek new markets and establish offices in alternative locations as a function of growth and competition. This interplay between 'us' and 'them' can be thus envisaged as a business system that needs to maintain closure under conditions of openness (Luhmann 1990).

5.4 Conclusions

Transformations of identity are inextricably linked to the evolution of a GSA relationship. The early yearning for doing high-technology development in India led to the establishment of ComSoft by Indian promoters who, after studying in the USA, wanted to return to India. Ghosh and Paul, the two senior GlobTel managers with Indian roots, harboured a strong desire to contribute to India's development through technology growth. In ComSoft, they saw a perfect vehicle for this aim. A hybridized cultural context drawing upon both the values of the Silicon Valley work environment and Indian values was consciously nurtured and constructed. This context both reflected and reinforced the desired values. A stable, creative and motivated ComSoft workforce was the key impetus to the growth of the GSA relationship in terms of volume and quality of work. The cultural context supported the realization of 'unleashing Indian creativity'. The growth of the relationship helped ComSoft develop confidence and ambition and in the process led to a realization of the limits of the GlobTel GSA. A gradual shift of focus to Japan came with ComSoft's realization of the limits of their existing image and their need for redefinition. In the meanwhile, the processes of growth of ComSoft and the Indian industry in general, along with the technological churn in the industry, raised the need for ComSoft to fashion a different cultural context from which a redefined image and organizational identity would emanate.

In a context of globalizing changes, as software and IT functions multiply, large companies like GlobTel find it a strategic necessity to outsource software services to distant locations and unfamiliar environments. However, developing successful inter-organizational relationships for intellectual and knowledge-intensive tasks involves concerns different from purely business and economic ones. In the present case, enrolling, retaining and developing services of bright programmers, technologically sophisticated managers and innovative entrepreneurs are fundamentally involved with issues of individual and collective identities and the implications and potential for their growth. Understanding such needs requires a dynamic conception of identity rather than one that is relatively static and stable. Changes in organizational identity and firm strategy may be understood in terms of situated actions of organizational members. Complexities of a self-referential pattern could often give rise to unintended consequences.

As discussed in chapter 4, GlobTel, like many transnational companies, tries to follow principles of 'standardization', seeking to make their products for development, management processes and practices in GlobTel and ComSoft 'nearly the same'. Creating 'sameness' fundamentally involves also matching in a complementary fashion the *identities* of the two firms. This matching is extremely complex as respective identities are guided and shaped in very different cultural contexts that are products of unique historical and social processes. We have expressed this cultural context through social structures relating to Indian society, global high-tech business, academic institutions and the firm. These structures have particular rules and resources that help shape identity and image. However, the fast-changing global situation requires a continual redefinition of identity to which actors must adapt and reframe their identities within the redefined cultural context.

Our analysis re-emphasizes the arguments of writers like Giddens and Castells about the central role of identity in the structuring of processes of globalization. GSAs are complex organizational relationships in a turbulent and dynamic global context involving international, inter-cultural aspects and broad implications for technology. Identity is significant to the shaping of managerial strategy. Changes at the multiple levels of the global, organizational and individual, are deeply implicated and, as Giddens (1990) emphasizes, are intensified in the current context of modernity and globalization. Globalization lies in constant conflict and tension with local particularities. Through processes of negotiations and learning, hybridized models of identity can be articulated and redefined.

BIBLIOGRAPHY

Albert, S. and Whetton, D. A. (1985). Organizational identity, in L. I. Cummings and B. M. Straw (eds.), *Research in Organizational Behavior*, Greenwich, CT: JAI Press, 263–95

Avison, D. E. and Myers, M. D. (1995). Information systems and anthropology: an anthropological perspective on IT and organizational culture, *Information Technology & People*, 8, 3, 43–56

Boland, R. J. (1996). Why shared meanings have no place in structuration theory: a reply to Scapens and Mackintosh, *Accounting, Organizations and Society*, 21, 7–8, 691–97

Burke, K. (1973). The rhetorical situation, in L. Thayer, *Ethical and Moral Issues*, London: Gordon & Breach

Castells, M. (1997). *The Power of Identity*, Oxford: Blackwell

(2001). Globalization and identity in the network society, *Prometheus*, March, pp. 4–18

Christensen, L. T. (1995). Buffering organizational identity in the marketing culture, *Organization Studies*, 16/4, 651–72

Covalski, M. A., Dirsmith, M. W., Heian, J. B. and Samuel, S. (1998). The calculated and the avowed: techniques of discipline and struggles over identity in six big public accounting firms, *Administrative Science Quarterly*, 43, 293–327

Dutton, J. E. and Dukerich, J. M. (1991). Keeping an eye on the mirror: the role of image and identity in organizational adaptation, *Academy of Management Journal*, 34, 517–54

Dutton, J. E., Dukerich, J. M. and Harquial, C. V. (1994). Organizational images and member identification, *Administration Science Quarterly*, 39, 239–63

Fiol, M. C. (1992). Managing culture as a competitive resource: an identity based view of sustainable competitive advantage, *Journal of Management*, 17, 1, 191–211

Giddens, A. (1984). *The Constitution of Society*, Cambridge: Polity Press

(1990). *The Consequences of Modernity*, Cambridge: Polity Press

(1991). *Modernity and Self-Identity*, Stanford: Stanford University Press

Gioia, D. A., Schultz, M. and Corley, M. (2000). Organizational identity, image, and adaptive instability, *Academy of Management Review*, 25, 1, 63–81

Godkin, M. A. (1980). Identity and place: clinical applications based on notions of rootedness and uprootedness, in A. Buttimer and D. Seaman (eds.), *The Human Experience of Space and Place*, London: Croom Helm, 73–85.

Hatch, M. J. (1997). Relations between organizational culture, identity and image, *European Journal of Marketing*, 31, 5–6, 356–65

Hofstede, G. (1980). *Culture's Consequences: International Differences in Work Related Values*, Bevercy Hills: Sage

Khare, A. (1999). Japanese and Indian work patterns: a study of contrasts, in H. S. R. Kao, D. Sinha and B. Wilpert (eds.), *Management and Cultural Values: The Indigenization of Organizations in Asia*, New Delhi: Sage, 121–38

Luhmann. N. (1990). *Essays on Self-Reference*, New York: Columbia University Press

Myers, M. and Tan, F. (2002). Beyond models of national culture in information systems research, *Journal of Global Information Management*, 10, 1, 24–33

Peters, T. and Waterman, R. (1982). *In Search of Excellence: Lessons from America's Best-Run Companies*, New York: Harper & Row

Robey, D. and Azevedo, A. (1994). Cultural analysis of the organizational consequences of information technology, *Accounting, Management and Information Technologies*, 4, 1, 23–37

Saha, A. (1992). Basic human nature in Indian tradition and its economic consequences, *International Journal of Sociology and Social Policy*, 13, 3, 1–76

Sahay, S. and Walsham, G. (1997). Social structure and managerial agency in India, *Organization Studies*, 18, 3, 415–44

Saxenian, A. L. (1996). *Regional Advantage: Culture and Competition in Silicon Valley and Route 128*, Cambridge, MA: Harvard University Press

Scott, G. S. and Lane V. R. (2000). A stakeholder approach to organizational identity, *Academy of Management Review*, 25, 1, 43–62

Sinha, J. B. P. and Sinha, D. (1990). Role of social values in Indian organizations, *International Journal of Psychology*, 25, 705–14

Smircich, L. (1983). Concepts of culture and organisational analysis, *Administrative Science Quarterly*, 28, 3, 339–58

Stimpert J. L., Gustafson, L.T. and Saranson, Y. (1998). Organizational identity within strategic management conversation, in D. A. Whetton, and P. C. Godfrey (eds.), *Identity in Organizations: Building Theory Through Conversations*, London: Sage

Walsham, G. (1998). IT and changing professional identity: micro-studies and macro-theory, *Journal of the American Society for Information Science*, 49, 12, 1081–9

(2001). *Making a World of Difference: IT in a Global Context*, Chichester: Wiley

Westrup, C., Jaghoub, S. A., Sayed, H. E. and Liu, W. (2001). Taking culture seriously: ICTs, culture and development, paper presented at *IFIP 9.4 Conference in Bangalore*, India (2002)

Whetton, D. A. and Godfrey, P. C. (1998). *Identity in Organizations: Building Theory Through Conversations*, London: Sage

6 The GlobTel–MCI relationship: the dialectics of space and place

6.1 The significance of space and place

Many contemporary writers have emphasized the fundamental role of the space–place distinction in contemporary life and globalization. For example, Giddens (1990) writes that in traditional societies, space and place largely coincided since social interactions occurred under conditions of 'presence'. In contemporary society, with increasing interactions between 'absent' others, space is separated from place, and activities are coordinated without necessary reference to the particularities of place. The experience of the 'here' and 'now' is tied to and contingent on actors and actions at a distance. Such discussions have helped to refocus the political role on the 'local' and to emphasize the tensions that arise in the interplay between the local and the global.

Space and place serve as powerful metaphors to understand the global and the local, respectively, and the tensions that arise when social practices play out simultaneously in the global and the local. Schultze and Boland (2000) describe place in terms of its association with the sense of boundedness, localness and particularity, as contrasted to space and its sense of universal, generalizable and the abstract. The distinction between place and space can be conceptualized with respect to the meanings that people ascribe to locations – physical or imagined. Spaces serve as containers or receptacles for places whose meanings are shaped by what one does in them. A place is a space to which meaning has been ascribed, and is psychologically meaningful as places help to ground identifications through the sharing of common symbols and experiences. In contrast, spaces represent physical areas with little associated meaning or intentionality.

A fundamental strategy that underlies globalization, and which serves as the guiding assumption of GSAs, is the notion that software development work can be broken up into modules, distributed to different centres around the world and coordinated through the use of ICTs to assemble the various pieces into the finished product. These different centres represent spaces, and serve as arenas for 'becoming', where one centre is not treated as different from the other, except on the economic criteria of costs and the availability of resources. This approach is in contrast to the situated perspective that work needs to be context-dependent and occurs in places, as arenas for 'being', where the local particularities are celebrated rather than suppressed. McDonald's outlets serve as a useful metaphor to understand this difference between space and place, since they are typically seen as arenas, similar to spaces where universalized and homogenized activities take place. These outlets are standardized to the detail of their physical appearance and food served. It is difficult to distinguish one outlet from another, even though they are located in different cities or countries (Leidner 1991). However, different people have varying sense of places and spaces. For the American tourist travelling in Rome a McDonald's outlet may represent a 'place' bringing back a bit of home; the same outlet to the Italians can represent an invasion of their place and eating culture.

GSW represents work arrangements that involve building software at a distance, outside the boundaries of organizations and countries, under the assumption that coordination is possible through the use of ICTs. The 'when' of these activities can be connected to the 'where' not through the mediation of place but by using technology. Castells (1996) writes that in contemporary capitalism, while organizations are located in places, and their components are place-dependent, their logic tends to be placeless and fundamentally based on the space of flows that characterize information networks. As networks in which firms operate become increasingly complex, they assume an increasing independence from the influences of the context of the locations they operate in. Fundamental to GSAs is thus the notion of geography, both physical and human, and how work plays out in them. However, notions of space and place when applied to GSA settings need to be extended from the traditional meanings ascribed to them by sociologists and human geographers to take into account the 'electronic shared spaces' (for example, through videoconference and email) in which a large proportion of GSW is carried out. Of interest then is how work practices that characterize GSW are structured across physical and electronic boundaries and how actors create and make use of distinctive conceptualizations of space, and the tensions that arise as a result.

The different physical and electronic domains take on space- or place-like character-istics depending on how individuals relate (or do not relate) to them and the practices that occur in these domains. They remain as spaces until individuals develop particular relations and identifications with them that transform them into places. There is an ongoing tension arising from this space–place interplay, such as when developers work in 'spaces' on projects that assume a 'place' kind of understanding. As individuals and organizations operate over space and place, there are varying dynamics that exert pres-sures simultaneously towards globalization and centralization on the one hand, and towards localization and regionalization on the other. This dynamic is confusing and conflicting, and creates both transition and displacement. Managers in GSAs constantly need to live and work with this disjuncture, in this 'in-between position', where they have to engage simultaneously with the global and the local.

Given the fundamental role of geography in GSAs – physical, human and electronic – the focus of this case is to analyse the relationships between GSW and different spatial forms including the human physical and electronic boundaries that are not physically defined but are articulated as movements in networks of social relations and understandings. This relationship between the social and geographic is redefined at different stages of development of the relationship. We analyse these changing forms in three stages of *initiation, growth* and *stabilization* of the GlobTel–MCI relationship.

6.2 Case narrative

Initiation (1991–1996)

MCI is part of one of the largest industrial groups in India with companies in a diverse range of industries: steel, automobiles, hotel chains, etc. The group turnover is around $5 billion. MCI resulted from the outcome of efforts of some pioneering senior executives in this group. They were one of the earliest in India to recognize the opportunities for software work globally and to establish an organization to translate the opportunity into a major business. MCI, from small beginnings, has grown significantly and is currently one of the largest software service providers in Asia. It has a network of more than fifty offices in the USA, Europe and Japan, with more than 10,000 employees. It had a turnover of around $400 million in 1999.

GlobTel's relationship with MCI started in 1991, described as a 'trialling exercise' with ten developers working in Mumbai. There was an initial sense of uncertainty and excitement about how the relationship would develop since MCI was traditionally a software consultancy firm with limited experience in telecommunications. Also, MCI had historically specialized in on-site work (derogatively referred to as 'body-shopping') and had relatively limited experience in offshore long-term development relationships, of the kind that a GSA entailed. In 1993, after a visit to North America, Mani (MCI's account manager for GlobTel) announced to his staff in India 'we need to become people in telecoms rather than in software'. Mani speculated on the three stages that MCI could go through in such a relationship: (1) to be a vendor on a fixed-price quotation basis; (2) increasingly to enter the core areas of switching; and (3) to become GlobTel's 'satellite lab' by taking on ownership of some GlobTel products.

At this stage, MCI identified two key challenges:

- How to manage the intricate relationship between technology and the issues that arise from working at a distance across time, space, and cultures.
- How to mesh the project culture of MCI with the product culture of GlobTel.

Mani said:

We are basically a project oriented company. We are not a product company, whereas GlobTel is a product company. So, I think there is a first challenge to try and mesh the people up, project culture versus the product culture. Project culture is typically a software consultancy or a matrix organization. We put a team together for a project. We do the project. After successfully completing the project, we hand over to the owner of the project and we disband the team till we start something else. Whereas in a product company like GlobTel, they expect a person to be there for a very very long time because it takes about 12 months to start appreciating the product.

Investments in technological infrastructure started to play an important material and symbolic role in the relationship. GlobTel set up a 9.6 kB baud link between India and North America without a video link, suggesting that GlobTel still wanted to 'test the waters' and keep MCI at arm's length. At this stage, GlobTel gave MCI mostly stand-alone work that was independent and involved minimum interaction with the company. As work started to evolve, GlobTel upgraded the telecommunications links to 128 KB including video lines, signalling to MCI an expectation of being 'close'. GlobTel established an expensive development environment in Mumbai that replicated their North American facilities. These infrastructural investments were reciprocated by MCI who established a GlobTel-dedicated lab in the suburbs of Mumbai where real estate was at a high premium. These mutual investments relied primarily on good faith, without a legally binding long-term commitment. They gave MCI the confidence of being 'included' in GlobTel's plans.

Expanded investment was matched with an increase in the number of developers in MCI's GlobTel group. They increased from 10 in 1991 to 70 in 1994 to about 300 in 1996, with an expectation of growth to 400 by the end of the year. GlobTel accounted for about $7–8 million of MCI's revenue, nearly 5 per cent of their overall international business operations, it was one of MCI's top 10 customers. Initially, MCI did 'lower-level' bug-fixing kind of projects that were relatively independent from the need to interact with GlobTel. Over time, there was increasing interdependence as MCI became involved in GlobTel's core areas of switching which required them to acquire domain-specific knowledge of telecommunications, of GlobTel products and at the management level to incorporate GlobTel's practices such as those related to internal reviews and quality checks. This interdependence created an increased need for 'proximity', and GlobTel upgraded their communication links from two to five–eight voice paths in 1994–5 that gave them more bandwidth to work with.

With this evolution of work, GlobTel became increasingly concerned with the loss of programmers (referred to as 'attrition'). After being trained and having acquired skills in the telecommunications domain, MCI programmers would seek more lucrative jobs, primarily in the USA. GlobTel tried to get MCI to institute measures to prevent this. MCI resisted what they called a 'Berlin Wall Approach' and instead saw attrition as something that could not be prevented but was a 'way of life' that needed to be managed. MCI desired autonomy to deal with the attrition problem according to norms existing in their organization. GlobTel and MCI subsequently both took a number of measures to deal with attrition. GlobTel located North American expatriates to MCI to try and 'install the GlobTel culture' and make MCI systems more 'objective'. MCI tried to create an internal 'GlobTel island' to prevent the movement of developers from GlobTel to other groups. MCI tightened their recruitment procedures and emphasized to potential recruits that they were being hired for telecommunications (not software) where the learning curve was much longer and more complex. Recruits were expected to give a verbal commitment to stay in the group for a minimum of 2 years, a requirement that did not extend to other parts of MCI.

Despite these initiatives, GlobTel's perception was that MCI's effort was inadequate. GlobTel was unhappy about MCI's resistance to their efforts to install new management systems, and viewed them as an 'outlier' in their international R&D relationships. Ram, a GlobTel expatriate of Indian origin, thought MCI was reluctant to take ownership of ongoing problems. GlobTel managers felt that MCI simultaneously relied on them for their vision, but also blamed them for 'shoving things down their throats'. MCI's perceived lack of cooperation and proactivity made Ram remark that GlobTel should critically re-examine whether the relationship was sustainable in the long run. Another GlobTel expatriate (of American origin) also was highly critical of MCI's management abilities and believed that the Indian managers were at least one or two levels lower in competence than their North American counterparts. She saw her primary mandate in India as trying to bridge this skills gap.

Growth in the relationship (1996–1998)

The period of growth was characterized by an increasing maturity in MCI and growing expertise in MCI both about GlobTel and telecommunications. In both India and North America, there was greater visibility to the relationship, which evolved from a 'contractor' to a 'partner' mode of working. Even with persisting concerns about attrition, micro-management and MCI's resistance to GlobTel's attempts to standardize, a number of key events took place in 1997–8 that gave a strong impetus to the relationship. A number of very senior GlobTel staff visited MCI and made a public commitment to support them. The GlobTel CEO, inaugurated the new MCI lab in Mumbai, and promised to clear obstacles at the strategic level to enable more effective joint work between MCI and SMTP (a GlobTel lab in Texas, USA). There was also a visit by Clement, the head of GlobTel switching operations in North America that accounted for $1 billion annual business, who announced that MCI was designated as the 'brother lab' for SMTP.

These public events signified a serious business commitment, moving MCI employees from a stage of 'regional-level' activities to an 'Olympic level'. Mani felt that 'A "brother-lab" implied a move beyond the stage of technical content to a stage of true partnership involving ownership and collaboration'. This heightened status to the relationship spurred both sides into action. While GlobTel was putting more money, people and executive time to make things happen, the MCI staff was actively debating the implications of being a 'brother lab' where they would not need to worry about project-to-project issues but on the overall strategic direction. They felt that the brother-lab status meant that GlobTel were affirming significant trust in them by inclusion in the 'family'.

Contributing to this significant growth was GlobTel's upgrading of the telecom link from 128 kB to 2 MB that would allow more interdependent work to take place. MCI felt that now GlobTel could not take decisions unilaterally by wearing their 'customer hat', but mutually in consultation with them. Mani gave an example of such a consultation:

I can tell you what a 2 MB link really means to the relationship. A senior person [from SMTP] had collected a lot of the evidence to substantiate his concern [against a particular MCI contractor]. A simplistic thing is to replace him, OK. Now you have this as an opportunity to turn this relationship into what it means because in the contractor model you do that, 'I don't like the contractor, please replace him'. But if there is ownership and a 2 MB link which indicates the closeness of the relationship, you would go and say 'I have the problem and how do we solve it?'. I cannot insist on my solution to the problem, because it is supposed to be our problem and I respect that.

During this period of growth, the problem of attrition became more significant. While both GlobTel and MCI were in agreement that attrition was a problem, they had contrasting views on how it should be dealt with. While GlobTel believed that attrition should be prevented, MCI saw it as part of life that had to be accepted and its consequences managed more effectively. MCI felt that GlobTel's 'Berlin Wall Approach' to prevent developers from leaving was a paradox. It curtailed developers' freedom while still providing them with the perception that they were free to move if they desired. Mani felt that attempts to prevent developers moving would be counterproductive and criticized GlobTel's standardized quantitative approach to solving even a human issue such as attrition:

The case in point is the standard problem of attrition and the GlobTel approach or the North American approach, I would not call it a GlobTel approach, is applied to it. I think the North American approach is to identify a need or a problem, collect data on that, put up a plan, put targets and track it to closure. So you take attrition, measure what it is, what is the attrition today, decide on a target. So supposing you say attrition today is 20 or 25 percent, you say I want to bring it down to 15 percent. So you say in the next six months we would bring it down from 25 to 20 percent, and then from 20 to 15 percent. There are certain problems, like the problem of attrition, that are far more complex, more local, where such a project oriented planning will not work.

The spectacular growth in the work demands created issues to which both sides responded by trying to increase proximity to the other. With increasing stakes in the relationship, GlobTel's managers felt the need for greater control through increased physical proximity. That required them to have a stronger physical presence in India. As work became more interdependent, MCI simultaneously felt the need to understand GlobTel's 'way of doing things' and sought greater physical proximity to them in North America. In seeking proximity, both sides expressed the need for a *place-based* rather than the *space-based* environment in which they had started.

GlobTel's 'GSODC' proposal: seeking proximity to India for closer control

GlobTel considered establishing their own Software Development Centre (GSODC) in India, as it would increase their first-hand and intimate understanding of the Indian context and help develop a sense of 'place'. For the centre to work, expatriates would need to be moved from North America to India at high relocation costs and salaries. This expense would be in opposition to GlobTel's initial objective of cost reduction through offshore work in India. The GSODC could potentially also allow GlobTel to sidestep some of the IP concerns arising from development through an external contractor like MCI. MCI resisted the GSODC proposal as they saw it as a direct competition to themselves. Development of new and core technologies would be handled by GSODC, and they be given the 'crumbs'. In the long run, this would stunt their movement up the value chain and they would also be more closely controlled from GlobTel.

MCI's need for proximity to understand GlobTel's 'way of doing things'

With increasing consolidation of MCI's work following the 'brother lab' announcement, a number of projects were transferred to them, signifying a move from Level 1 to Level 3 work. A senior MCI manager described Level 1 work as 'the easy, by-the-kilogram-kind of stuff' (for example, bug-fixing and regression-testing projects), in which developers became productive and could deliver within 2–3 months. Level 2 work involved adding new features to existing software. Level 3 work involved responsibilities for the entire product, including developing enhancements and responding to customer complaints at service levels similar to GlobTel. The move from Level 1 to Level 3 implied a massive leap of faith for both sides, which Mani described in the following manner:

> The buck-stops-with-you kind of thing. That is what I call throwing your hat over the trapeze, Level 3 is the highest level. You are literally an equal partner of GlobTel.

The move to Level 3 implied increased expectations and responsibilities for MCI. It created the need for them to better understand GlobTel's 'way of doing things'. Mani theorized that GSO relationships involved the transfer of knowledge of products (for example, specifications of DMS switches), of processes (for example, of quality control) and of practices (referring to local ways of doing different kinds of tasks). He felt that while GlobTel had done very well on the first two, they had generally overlooked practices, resulting in misunderstandings and often conflict between the two sides. MCI's managers were perplexed and frustrated as they felt that, despite meeting targets, GlobTel remained unsatisfied. While sitting in some GlobTel's internal meetings in North America, Mani realized the reason for this lack of satisfaction. He saw that the templates used by the North American managers to report results from projects were quite different from those used by the MCI managers. While the Indians presented only hard details of the project (deliverables, time deadlines, etc.) the GlobTel staff also gave details on intangibles such as the satisfaction of users, etc. The Indians did not report such intangibles since, Mani felt, they were rooted in situated practices that could not be transmitted in training programmes as easily as products and processes. To be able to do so in the future, Mani proposed to GlobTel that:

> [We] ... must have a management entity that is distinct from GlobTel, [our] own entity in North America. Let us concede the other way that if MCI is incapable of doing it even in North America they are definitely not going to do it in Mumbai; you might as well give up. If I cannot run it out of North

America, I cannot run it out of India anyway. I have no excuses such as lack of knowledge or practices. I have no such excuses anymore . . .

MCI believed that the increased costs at GlobTel by higher on-site presence would be marginal, and the total would still be less than North American costs. More importantly, did the managers with this enhanced proximity gain an increased 'comfort level', especially in the prevailing context of churn and turbulence in the industry? GlobTel resisted MCI's proposal to increase their on-site presence to 30 per cent, and on the contrary, it was reduced from 15 to 5 and was later renegotiated to 10 per cent. Against the background of these ongoing tensions of proximity (place) and distance (space), we describe the phase of stabilization of the relationship.

6.3 Stabilization of the relationship and residual tensions (1999–2000)

The challenge to the economic viability of the relationship was accompanied with a growing maturity where both sides showed an increased understanding of the other's problems. Mani described how the condition of both sides speaking 'past each other' (as was evident in the earlier phases of the relationship) was changing:

So we are being forced to think through it. Previously we were saying you are not listening, so people were getting very emotional. Now I think the wrinkles on everybody's forehead mean that people will now listen and I think we will also do some more homework because now we think people will actually listen to us.

Mani gave an example of attrition where there was now agreement concerning the expectations of the other, even though there was a lack of agreement on the substantive contents:

In the last year, I think a lot of people have come to terms, understood what are the gravity or complexities of attrition. So now, although they are very disappointed in the sense that they thought if we did some nice initiatives, made some recognition measures and so on, attrition would significantly go down, they are far more aware that the complexity of the problem is much deeper. Now we understand that attrition is something we have to manage not solve. But I would say it is probably more realistic. We can get into a more meaningful dialogue with people today than we could last year.

This stage of stabilization, as reflected in the common framework for dialogue, was accompanied by other tensions, including the 'right-angle turn' that raised the question of whether MCI should support this turn or help maintain the legacy products. The latter option would help free GlobTel's expensive resources to work on 'webtone' product development, making budgets for dialtone systems a 'shrinking pie'. A sense of stability came as the legacy option was selected. In 1999, many senior staff from both sides who had been responsible for establishing the relationship either resigned or retired (for example, Paul and Ghosh from GlobTel and Mani from MCI). For MCI, 1999 represented an all-time low in the relationship in terms of growth. Ashok (Mani's replacement) described MCI's management as being unhappy with this, especially given the manner in which the rest of the software industry and other parts of MCI were growing. In the context of the 'shrinking pie', the potential for future growth of the relationship seemed bleak. Stabilization thus seemed temporary, and fraught with its own internal contradictions and tensions.

A summary of the key events and their interpretation is provided in table 6.1, which is then followed by the case analysis.

6.4 Case analysis: the perspective of space and place

The case analysis is presented at three levels of abstraction:
- At the first level, the aspects of space and place are analysed in the three stages of the relationship
- At the second level, we discuss the dialectical nature of the interplay between space and place
- At the third level, the implications of this dialectical interplay are examined in relation to the process of growth of the GSA relationship.

Initiation

In the initial stages of the relationship, MCI and India represented a 'space' for GlobTel, not different from the other labs they had worldwide. MCI was seen as an arena for 'becoming', a location that could provide reduced costs for software development and improved access to skilled human resources. It was assumed that the local particularities of India could be superseded by the shared electronic spaces and MCI could become just another one of GlobTel's many development centres. From the start, GlobTel tried to standardize their operations so that their managers could seamlessly work across these settings as if still working within the GlobTel framework. Seen as a universal location for 'becoming', in which universalized practices could apply, GlobTel introduced a number of their globally standardized approaches – for example, the model of technology transfer and metrics for measuring lab productivity.

Understanding the sense of space is important because it helped to shape the expectations of senior GlobTel staff, and to provide them with a frame of reference to evaluate subsequent events. The development environment was replicated in the MCI lab with the inside of the lab resembling a modern North American office with videoconferencing and telecom links that allowed people in North America to ring MCI staff in India (and vice versa) as if making a local call within their own building. In trying to create a 'place' for their managers similar to that in North America, GlobTel assumed that MCI and Mumbai were a 'space' where local particularities could be superseded by the GlobTel standardized environments. Managers from both sides had to operate constantly and

Table 6.1 Time line of key events in the GlobTel–MCI case study

Time period	Key events	Our interpretation of key events	Basis for interpretation
Initiation phase (1991–6)	Initiation: 1991 Telecom links rise from 9.6 kB to 128 kB Work evolves slowly from level 1 to level 2 MCI establishes dedicated lab for GlobTel work in Mumbai GlobTel invests in replicating Ottawa development environment in Mumbai GlobTel enhances voice paths from 2 to 5–8 to support independent work Number of developers in MCI working on GlobTel projects increases from 10 to 70 and then to 300 GlobTel is one of MCI's top 10 customers and contributes 5 per cent of their overall international revenues GlobTel locates expatriates in India Attrition is a key issue	Gradual evolution of level and quantum of work Growing investment in infrastructure from both sides signals building of long-term commitment GlobTel's attempts to micro-manage resisted by MCI	Historical reconstruction
Growth phase (1997–8)	Senior GlobTel executives visit Mumbai MCI is designated as GlobTel's 'brother lab' GlobTel proposes to establish own development centre in India (GSODC) MCI proposes to increase on-site presence in North America Work evolves to Level 3, involving the transfer of ownership Growth in number of programmers – around 400 Relationship reaches a high in terms of MCI revenue Attrition seen as a burning issue	MCI see themselves as being 'included in the GlobTel family' MCI resist GSODC initiative as it will stunt their growth GlobTel reject MCI initiative as it will increase costs Increased responsibility and expectations of MCI GlobTel finds it hard to release responsibility of ownership and control	'Real time' interviewing
Stabilization phase (1999–2000)	'Right-angle turn' for GlobTel to respond to the Internet challenge GlobTel decides to use MCI to support legacy systems rather than the 'right-angle turn' Relationship stagnates in terms of revenues and number of programmers, pessimistic outlook Stabilization, though not agreement, in understanding of expectations of the other	MCI left holding the legacy systems Disappointment as a result GlobTel seeks new possibilities in India	'Real time' interviewing

simultaneously in place- and space-based environments as they moved between India, North America and the electronic domains.

During the early stages, the partners were still getting to know each other, and the relationship was still relatively 'distant'. There was uncertainty on both sides. The Level 1 work that was given to MCI did not require 'shared' spaces. The degree of shared spaces was minimal (9.6 kB baud), and this lower bandwidth in turn implied that lower-level work was given. More numerous visits of GlobTel's staff to India and vice versa helped to increase mutual familiarity. GlobTel's increasing investments in telecommunications infrastructure started to enhance the shared spaces, which further contributed to up-grading the quantum and level of work. These investments were interpreted by MCI as indicative of GlobTel's greater long-term commitment and desire to make India a 'place'. MCI reciprocated the investments by dedicating a large exclusive building in Mumbai for GlobTel work that would serve as a 'place' for both the Indians and the North Americans – for the Indians, because it was an exclusive 'island' unique in the MCI structure, and for GlobTel because the physical and symbolical environment replicated their home office.

The increasing presence of GlobTel expatriates started to create tensions as MCI tried to protect their autonomy. In turn, GlobTel tried to legitimize their presence by pointing to various problems in MCI's management that created the need for physical supervision and correction, such as the issue of attrition. Expatriate visits were not welcomed by MCI because they were associated with control rather than with learning and developing a 'place-like' understanding. Both distance and proximity helped to shape interactions in the initiation stage. To counter the geographical realities of distance GlobTel introduced the proximity-enabling characteristics of telecommunication links and the presence of expatriates, The interplay between proximity and distance helped to shape the nature of work processes (for example, the kind of projects) and interactions (for example, the need for control and autonomy). Technology played a key role in shaping these dynamics as it provided the potential to bridge the geographical and temporal realities. It also inscribed symbolic meanings of commitment and proximity. These helped to shape the sense of space and place which actors experienced in the physical and electronic domains.

Growth

The period of growth was an interesting and dynamic time for the relationship of MCI and GlobTel. It was characterized by rapid growth in the number of developers, increasing levels of work and significant exchange of people from both sides. A turning point was the visit of very senior GlobTel staff members to India and their public announcements of commitment. MCI interpreted this not as a desire for control but as an expression of long-term commitment, of being included in the GlobTel 'family'. There was a 'merging of places' through these visits, as MCI became a 'brother lab'.

The merging of MCI and GlobTel had direct implications for the nature of work being done by MCI, because the focus shifted to transferring ownership to the MCI lab.

It would then have overall responsibility to support and enhance a piece of software or a product feature. However, to fulfil this responsibility, MCI needed more knowledge about how GlobTel did things, not through shared spaces but through physical 'place-like' proximity. But this physical proximity implied increased costs and was thus unacceptable to GlobTel. Simultaneously, GlobTel tried to establish their 'place' through a development centre in India, which the Indian partners resisted as it was seen as detrimental to their business interests and autonomy. While the 'brother lab' status dissolved the sense of space and distance, its implementation presupposed a need for place for both sides but in opposite directions. Both MCI and GlobTel felt the limits of electronic space, MCI felt these limits in accepting the responsibility of full ownership and GlobTel in letting go of this responsibility freely.

The 'space' and 'place' metaphors, when integrated with the geographical and material realities, help understanding of some of the tensions in this phase. MCI's desire for geographical proximity to help develop a 'comfort level' was balanced by GlobTel's cost considerations. Geographical proximity for GlobTel implied potential of loss of business for MCI. Both sides wanted simultaneously to increase proximity and 'maintain their distance'.

Stabilization

Stabilization was attained with both sides apparently understanding the place of the other and the activities that were legitimate in it. On the issue of attrition, for example, through negotiations over time GlobTel came to respect this as a local MCI problem. While both sides realized the gravity of the problem, they also came to respect the other's position on it.

While space and place are generative principles drawn for the analysis, they are not reductionist, as they are examined in relation to other events such as the 'right-angle turn'. MCI was largely excluded from this new development on grounds of increased need for expertise and proximity. GlobTel's legacy work found its 'place' in MCI, although reluctantly. This sense of stabilization appears temporary owing to the inherent tensions and contradictions that could potentially alter the dynamics of the relationship in the future.

After this first-level analysis based on the concepts of space and place, we present another level of analysis of the dialectical nature of the space–place interaction. This is then followed by an analysis of the implications of the dialectical nature of this interplay in the process of growth of the relationship.

6.5 The nature of space–place analysis: a dialectical perspective

The Greek term 'dialectic,' derived from 'dialogue' between equal partners, implies a unity of opposites. In a dialectical relation, each of the opposed 'poles' is incomplete without its own 'other' and disappears when abstractly isolated. Hegel describes the concept of the master to imply the opposite of such a concept in the servant, the one who is mastered. Without the idea of one, we cannot form the idea of the other, although each idea in itself is the contradiction or negation of the other. Similarly, in GSAs the idea of space (and distance) cannot be understood without its opposing category of place (and proximity). This dialectical interplay between place and space over time helps to develop a richer understanding of the process of evolution of the GSA.

An important question concerns what constitutes *dialectical reasoning*. One view is that it is an ontological stance to represent a view of knowledge, and the second is that it is a convenient epistemological approach to access reality. Harvey (1996) considers this debate as spurious, since it involves both epistemology and ontology. Dialectics are concerned with how we *abstract understanding* from a phenomenon that we encounter in everyday life. There are also different versions of dialectics, one being from the strong version that is associated with the deterministic sequence of thesis, antithesis and synthesis. The alternative is the largely interpretive view that emphasizes the mental. This strong view has many critics because of its essentialist nature and deterministic implications.

Harvey argues that criticisms of the dialectic approach can be avoided by focusing on the *mental and representational aspects*, where dialectics are not viewed as a form of logic but in an interpretative sense to represent a flow of arguments and practices that we use to interrogate and describe a phenomenon. There is an ontological stance in that the understanding of processes and flows is emphasized over the analysis of elements and things. The epistemological stance inverts the emphasis in that 'we get to understand processes by looking either at the attributes of what appear to us in the first instance to be self-evident things or at the relations between them' (Harvey 1996, p. 49). This epistemological emphasis is again reversed when it comes to formulating abstractions, concepts, and theories about the phenomenon. Harvey argues that this process of inversion transforms the self-evident world of things into a more confusing set of relations and flows that are manifest as things. Harvey gives the example of capital which is conceived as a stock of assets at a particular value (as a set of things) which is constituted in processes of capital production and exchange (flows). A dialectical reasoning rests on understanding how things and processes are constituted in each other.

We adopt the dialectical approach to reasoning in the sense described by Harvey: to represent both an ontological stance and a method. We use this method to access the nature of processes inherent in a GSA. We adapt this process of reasoning to further analyse the nature of the interplay between space and place, and how that relates to the GSA process. This analysis is conducted by focusing on three key dialectical principles of *totality*, *change* and *contradiction* (Rees 1998):

- *Totality* insists that the various seemingly separate elements in the world must be related to one another

- *Change* suggests that when we bring together these different parts, their meaning is transformed in a qualitative way, this arises not from external forces, but from a universal 'changefulness' located within the phenomenon
- *Contradiction* implies a union of two or more internally related processes that simultaneously support and yet undermine one another.

These dynamics are both integral to, and yet destructive of, the processes themselves.

The principle of *totality* guides us to an analytical focus on the linkage between spatial forms and material practices, and the unfolding of the linkage over time. This helps to include the entire network of actors, their social practices and and technologies, that influence both the material and symbolic domains. Within this conceptualized totality, the space–place dialectic serves as generative principle on which the evolution of the GSA is analysed. GlobTel initially treated India as a space, and also simultaneously sought to develop understandings of local particularities. MCI attempted to keep autonomy over their place (reflected in their resistance to expatriates), but at the same time needed to operate in space-like environments to ensure cost effective coordination.

A focus on the material–social linkage helps one to understand the expectations of actors and how these are shaped by a 'space'-kind of understanding. GlobTel believes in the power of telecommunications links to transcend the physical and the cultural distance, and reduce costs and time to market. Being a large MNC with worldwide operations, GlobTel deals with the uncertainty of India as it would with other countries and spaces: by standardizing activities and locations so that their managers can seamlessly work across different settings as if operating within a uniform GlobTel framework. The development environment in the MCI lab replicates the North American setting, including the air-conditioned lab and similar telecommunications links. Through standardization of physical and symbolic environments, GlobTel tries to create a 'place' for their managers, superseding MCI's 'place'-like local particularities.

The principle of *change* emphasizes internal 'changefulness' and the associated sense of permanent instability. The relationship between MCI and GlobTel was in a constant state of change, not merely an external displacement but a qualitative alteration in the nature of work being done over time. In the initial stages, MCI did lower-level projects in 'local spaces'. While MCI experienced autonomy with independent work, they simultaneously aspired to be 'closer' to GlobTel so they could work on new and exciting technologies. Over time, GlobTel's confidence in MCI grew and work evolved from Level 1 to Level 2. To support that growth, GlobTel created more electronic 'shared' spaces. Simultaneously with increased autonomy for MCI, GlobTel established various controls to enable integration of local work with global operations. The 'brother lab' announcement brought a major change, as GlobTel signalled their desire to transfer ownership.

The place itself, both in India and globally, has changed from 1991 to 1999 and these changes influenced the course of the relationship. The number of MNCs operating out of India in 1999 was significantly greater than in 1991, creating different competitive pressures and a higher level of telecommunications-related domain expertise in the industry. With changes in knowledge levels, the expectations and aspirations of both the organization and the individual programmers were redefined, creating new challenges and opportunities in the relationship. The technological aspirations of the Indians were not initially a consideration, but later on they affected the value of the relationship in the view of the MCI developers. Technology plays a key role in shaping these dynamics, representing both a sense of exclusion–inclusion and autonomy–control, materially and symbolically.

Change is inherent in the process of growth in any relationship. A dialectical analysis leads us to understand the constant qualitative change that is internal rather than external to the relationship between MCI and GlobTel. Initially, cost reduction becomes a major consideration, later on, it is not a primary concern. Initially, the Indians had limited domain knowledge of telecommunications and GSW; with time, as they acquired and developed such understanding, their expectations changed. Knowledge thus became a key driver of expectations and the shaping of further changes.

Contradiction in a dialectical context helps to explain rather than describe change. Space and place simultaneously support and undermine one another. It is only through working at a distance, in space, that the need for proximity, in place, arises. With increased proximity, actors feel autonomy being threatened and express the need for 'space'. Processes over space and place thus cannot be understood without each other, and yet their joint influence generates conflicts among associated structures and processes. There are inherent contradictions between the intentions of GlobTel managers who want to minimize on-site work in North America and the Indians who see the relationship as an opportunity to get closer to the North American marketplace. As these contradictory needs are better understood, some are recognized to be inherently irreconcilable. For example, the Indians' aspiration to be involved in new technology work was contradictory to GlobTel's view of India as primarily a 'space' for legacy systems support. GlobTel's fundamental management strategy of standardization is contradictory to MCI's need to maintain their Indian identity while working in a global environment. For the success of the relationship, MCI had to accept standardization to a certain degree, but the level and extent to which they did so was contentious and was constantly negotiated and contested.

The evolution of the relationship helps both to support and undermine the relationship. On one hand, as MCI developed expertise in telecommunications, they bid for contracts in other locations such as Japan and Korea. From GlobTel's perspective, this had a negative effect on the relationship. On the other hand, with increased familiarity of business in India, GlobTel gained a sense of 'place' in India, and built up confidence to establish their centre and to bid for Indian and Asian markets more aggressively. While Japan provided different kinds of technological possibilities to MCI, GlobTel could potentially use their Indian centre for doing new technology work on their own instead of outsourcing it to the Indian firms. This, of course, negatively impacted to the relationship when viewed from the MCI perspective.

As people work in various physical and electronic domains, they experience varying and contradictory feelings of inclusion and exclusion. Operating independently in the local domain, the Indians experienced a sense of being 'hands-off' and excluded. Increased proximity in shared spaces was contradictory to the Indians' needs to have control over their local place. The initial exciting prospect that GlobTel offered for new development was slowly changed over time and MCI itself resigned to lower-end legacy work. This state of resignation is unlikely to endure because of its own internal contradictions and tensions. MCI management, unhappy with the low-growth situation, are expected to reorient their strategy away from legacy systems in the future.

Taken together, the three dialectical principles of totality, change and contradiction help to develop insights into the micro-level dynamics of how practices over space and place play out. We next address the broader research question of how the analytical lens of the dialectics of space and place can provide insights into the process of the GSA relationship.

6.6 The dialectics of space and place and the process of GSA evolution

A fundamental requirement for understanding the GSA process is the need to focus on space and place. We argue that space and place should be given the status of 'root metaphors' (Pepper 1942) in which various GSA-related activities unfold. Pepper describes 'root metaphors' as concepts that people may not be normally aware of, but which they unconsciously draw upon as a means of 'seeing' the world and ascribing meanings to things. GSA arrangements involve building software at a distance, outside the boundaries of organizations and countries. Fundamental to understanding GSA is thus the notion of geography, both physical and human. We have analysed these elements of geography and their relationship with GSA practices through the concepts of space and place. Space and place are fundamental in shaping various decisions, including whether or not to enter into outsourcing relationships, what kind of projects should be outsourced and what kind of control systems will be required.

We have extended the notion of space and place as traditionally used by human geographers and sociologists to include 'electronic shared spaces'. Electronic spaces, enabled by videoconference and email, for example, are fundamental to the analysis of the process, as this is where a large proportion of GSW is carried out. A shared domain provides a flexible bridge between diverse global places within a 'space'-like framework. This assumes 'place'-like characteristics unique to the 'now' state of the relationship. Different degrees of proximity and distance characterize this 'shared place', depending on the state of development of the relationship. Technological parameters such as bandwidth reflect the frequency of interaction. Use of audio or videoconferencing characterizes an in-between state. A flexible third state of 'virtual presence' may be invoked to characterize relationships in addition to those of physical presence and absence discussed extensively in previous literature (e.g. Giddens 1994).

However, the case brings out the limitations of this 'virtual presence'. While enabling better understanding of some aspects of the relationship, it does not help understanding of others. Aspects relating to the larger context that give rise to attrition or tensions arising from strategic ambitions never figure in the shared space. These remain place-based to Mumbai or North America, requiring GlobTel, despite all its expertise and trust in telecommunications, to depute Indian-origin expatriates to India to help understand complex issues.

Meanings which actors ascribe to these spaces (for example, commitment), and the material features of these spaces themselves (for example, bandwidth) have important implications for the structuring of work. The local, global and shared domains take on 'space'- or 'place'-like characteristics depending on how individuals relate (or not) to these domains and what activities they engage in. These domains remain as spaces until actors develop the familiarity that transforms them into places. In the complex context of GSA relationships, such needs generate a dialectical oscillation between space and place. There are varying tensions at different stages of the relationship. Initially, there is the inclusion–exclusion tension because of the geographical realities of distance and the historical realities of the GSAs' inexperience. This distance has to be bridged for the relationship to develop and for both sides to see value in continuing. Bridging takes place in many different ways, including the use of technology, the increase of physical presence and the standardization of work processes. These mechanisms have both material and symbolic implications for the relationship. They permit higher-level work, signify greater commitment, change management expectations and introduce new tensions.

With the bridging of some of these physical and cultural distances, there is a significant development in the relationship in the growth phase. The growth expected to come through a 'merging of places', however, is fraught with key tensions between autonomy–control and explicit–implicit knowledge. For the relationship to grow, GlobTel needed to release responsibility and MCI to accept it. They both recognize the need but have different views on how this should be done, and how well the other is doing. GlobTel tried to exercise greater control through increased physical proximity, which came only at the expense of MCI's autonomy. MCI tried to reclaim and renegotiate this threat to autonomy by seeking physical proximity in North America, a move contested owing to cost concerns.

External events driven by technological changes led to both a stabilization and desta-bilization of the relationship. Although these external events led to stabilization with MCI being left with the legacy systems, this situation was not sustainable in the long run. This supports Slack and Williams' (2000) argument that the dialectics of space and place need to be studied in the context of the contradictory implications of ICTs. These contradictory dynamics illustrate the largely indeterminate nature of GSA relation-ships, and help to argue against the adoption of purely deterministic models of growth. Different configurations of the relationship can emerge as the space–place dynamics interface with such external events. The three dialectical principles of totality, change and contradiction sensitize us to the processes through which alternative forms may emerge. Through this analysis, we have extended Harvey's (1996) agenda to go beyond the invocation of space and place as convenient metaphors and to integrate them more meaningfully with material and geographical realities. Such integration, Harvey argues, 'not only has a transformative effect upon the domain of theory, but also opens up a terrain of political possibilities' (1996: 47). We have opened up political possibilities in various domains of GSW, including around standardization.

In chapter 4, we discussed the Witech case with a focus on standardization. The space–place analysis helps to develop additional insights into the limits of standardization that can be achieved in GSW, and the reasons for those limitations. GlobTel set up the arrangement in India on the assumption, that with telecommunication links and standardized practices, the problem of distance could be overcome. Their motivations for working in India were the potential resource and cost advantages, as well as the available bandwidth. These motivations were reflective of broader discourses on globalization and modernity that were biased against place and local particularity (Dirlik 1998). Our case has emphasized that for a variety of reasons, 'place matters' and can never be eliminated. GSW requires both the immediacy of face-to-face co-presence and action at a distance, requirements that are inherently dialectical in nature. Space-based work conflicts with the managers; and developers' 'compulsions for proximity' (Boden and Molotch 1994), their need for co-location with their development partners. Human beings have a fundamental and enduring necessity for co-present communication. In distributed software development, co-presence is particularly useful as it is 'thick' with information and can potentially deliver far more context to developers than can communications through technology. However, as our case highlights, there are liabilities of co-presence arising from risks of being 'micro-managed'.

Co-present interaction contributes to the development of solidarity and articulation of trust in the manner and timing in which actors place talk and gestures in a conversation. Solidarity is also expressed when participants emote together – for example, laughing together at the 'right' time. The flexibility of co-presence allows trivial and apparently unimportant talk to occur on occasions otherwise dedicated to prescribed topics. Co-present interactions imply an expression of commitment, because it requires participants to set aside not only a specified time but also a shared space, at the cost of constraining other activities. The physical situation of co-presence allows participants to deal with circumstantial contingencies, especially during sensitive and uncertain situations like design where there is no set script or standing recipe for arriving at an outcome, and things must be worked out along the way. Co-present interaction also supports the development of tacit knowledge, which describes our ability to perform skills without being able to articulate how we do them. Since such knowledge cannot be formalized and passed on, by its very nature, tacit knowledge is best developed in co-presence where the learner watches the expert and mentally assimilates how he/she makes things work.

In the context of distributed work, Maznevski and Chudoba (2000) describe the importance of having initial co-present meetings to establish effective temporal rhythms like 'heart beats', which support subsequent stages of global virtual teamwork. The need of individuals for 'place' kind of understanding is in conflict with the organization's 'global strategy' that involves close coordination of worldwide operations with central control, and is supported by standard product designs (for example, telephone instruments), management practices and technical expertise. GlobTel coordinates their operations by modelling particular knowledge systems and transferring them to their development partners worldwide, including MCI and Witech. The evolution process starts with independent work and moves to technical standardization (setting up the networks) and then to management systems (systems of HR appraisal) and ultimately to the organizational level where MCI becomes a part of the GlobTel 'family'.

Standardization processes are attempts to create 'space' kind of environments that enable homogenized operations and more efficient coordination. Various ICTs help to standardize GSW, and also provide the potential to let local practices flourish. Thus spaces are created as videoconferencing allows projects in India and Russia to be treated with similar and homogeneous project management structures. At the same time, this technology helps people in North America feel as if they are still within their organization even though they are interacting with an Indian or a Russian company. While the availability of new technology helps to develop standardized spaces, people tend to subvert or adapt these universalizing efforts to their environments to maintain local practices. Standardization attempts reflect a contradiction of highly local and immediate actions with distant, global moments and practices. In GSAs, this contradiction comes with the needs of the place-based work which software development entails, and the geographical imperatives of distance implied by globalization. These tensions make popular expectations of technology such as 'geography into history' or 'follow the sun' untenable to a significant extent.

While there is always this tension between standardization and local realities, some amount of standardization is desirable for purposes of coordination and control. A software package being developed by programmers in three countries (say India, Russia and Canada) needs to be coordinated to ensure that the development is integrated and the product works together in totality. This requires a degree of standardization of the development process, system of documentation, etc. Designing systems that are totally locally sensitive is also prohibitively expensive. There is the need for developing a pragmatic balance that strikes a blend between the need to be sensitive to local context and the need to develop standardized solutions universally (Rolland and Monteiro 2002). The space–place analytical lens can help support this quest for balance by emphasizing the potential for breakdowns. Both in material and symbolic terms, we can examine how standardization attempts threaten both organizations' and peoples' sense of physical place or need for autonomy, and help to draw inferences about the domain and extent of standardization. While it may be easy to standardize processes relating to transfer of knowledge of products because of the more universal and 'space'-like characteristics associated with processes, it is inherently problematic to homogenize management practices and behaviour that are strongly place-based.

BIBLIOGRAPHY

Boden, D. and Molotch, H. L. (1994). The compulsions of proximity, in R. Friedland and D. Boden (eds.), *NowHere Space, Time and Modernity*, Berkeley: University of California Press, 257–86

Castells, M. (1996). The network society, in M. Castells, *The Rise of the Network Society*, Oxford: Blackwell

Dirlik, A. (1998). Globalism and the politics of place, *The Society for International Development*, 41, 2, 7–13

Giddens, A. (1990). *The Consequences of Modernity*, Cambridge: Polity Press

(1994). 'Foreword', in R. Friedland and D. Boden (eds.), *NowHere Space, Time and Modernity*, Berkeley: University of California Press, xi–xiii

Harvey, D. (1996). *Justice, Nature and the Geography of Difference*, Oxford: Blackwell

Leidner, H. (1991). *Fast Food: Fast Talk*, Berkeley: University of California Press

Maznevski, M. L. and Chudoba, K. M. (2000). Bridging space over time: global virtual team dynamics and effectiveness, *Organization Science*, 11, 5, 473–92

Pepper, S. C. (1942). *World Hypothesis: A Study in Evidence*, Berkeley and Los Angeles: University of California Press

Rees, J. (1998). *The Algebra of Revolution: The Dialectic and the Classical Marxist Tradition*, London and New York: Routledge

Rolland, K. H. and Monteiro, E. (2002). Balancing the local and the global in infrastructural information systems, *The Information Society Journal*, 18, 2, 87–100

Schultze, U. and Boland, R. J. (2000). Place, space, and knowledge work: a study of outsourced computer systems administrators, *Accounting, Management and Information Technologies*, 10, 187–219

Slack, R. S. and Williams, R. A. (2000). The dialectics of place and space, *New Media & Society*, 22, 313–34

7 Managing the knowledge transfer process: the case of Sierra and its Indian subsidiary

7.1 GSW: a knowledge perspective

This chapter introduces the importance of taking a knowledge perspective to understand the process of growth of a GSA. Software development in general is an example of knowledge-intensive work. When taken in the context of global separation, the knowledge intensity is magnified many times over. In addition to domain-specific product knowledge, it requires a deep understanding of the multiple local contexts involved and the ongoing and changing development requirements. Many of these requirements remain undocumented and tacitly held by individuals and groups.

The significance of the knowledge transfer process can be analysed at multiple levels: institutions, project teams and individuals. Institutional reflexivity, which Giddens (1990) has described to be a defining feature of globalization, is particularly emphasized in knowledge-intensive software development work. Continuously and reflexively, firms must monitor new knowledge about technologies, organizations and markets. They must make changes in their own processes as the situation demands. The need for continuous reflexive action is significant because of the speed at which new knowledge about software is being generated and the tremendous interconnectivity between global systems (including electronic) that facilitates the spread of new knowledge from one part of the world to others.

The knowledge perspective is equally significant at the level of project teams and the individuals in the teams. Most software development firms comprise multiple teams working with different clients, engaging in different technologies and serving various markets and products. There is thus a significant amount of *localized knowledge* which is confined to these project teams. Access of this knowledge by the rest of the organization is often limited for a number of reasons, including the informal and tacit nature of such knowledge. The activity of software development is dependent largely on the level of skills and expertise of individuals. Hence attrition is one of the biggest challenges to effective management of GSA relationships. Movement of developers and their individual preferences for particular hardware and software development platforms affect the status of certain types of knowledge availability. For example in 1998, some Bombay software programmers we interviewed perceived Powerbuilder to be a more lucrative platform to work on as compared to Oracle. Their reason was that Powerbuilder made possible employment opportunities in North America. As a result, many Indian software outsourcing firms and their foreign clients who utilized Oracle platforms were left facing a shortage of appropriately skilled staff, and the problem of retaining them over time.

The knowledge perspective in GSA needs to be integrated with the temporal dimension. Organizations have different demands for information and knowledge at various stages of their evolution. They require different strategies to deal with the different demands over time. In the cases discussed so far, we have pointed to the different phases of a relationship from initiation, to growth and varying degrees of stabilization (or lack thereof). Many organizations, when establishing a GSA, start with small projects where a team of developers from the vendor company go on-site to the client's premises (Nicholson 1999; Nicholson and Sahay 2001). Most often, a structured approach to development is adopted to understand the various products, processes, practices and notations through which 'knowledge' becomes readable and usable by those involved in the software development process (Telioglu and Wagner 1999). Products could include programming languages and application tools. Technical processes relate to development and project management methodologies and management includes a quality process such as ISO 9000 and the Capability Maturity Model (CMM). Practices involve culturally accepted norms such as approaches to learning, problem solving and communication. Notations include specifications, flow charts and other similar inscriptions.

As the developers from both sides of the GSA sit in conditions of co-location during the initial stages, it can be expected that knowledge on various facets of the process can be 'transferred'. An increased proportion of the development tasks can then be moved offshore where the costs of development are cheaper. This movement to offshore development is facilitated by the creation of trust arising from the early experiences and interactions which had supported the building of personal relationships ('I trust X because I know him/her and she/he won't let me down'). Trust is also based on confidence that deadlines and quality standards will be consistently met when committed to, which is backed up by structural controls and contracts.

Managing the balance between offshore and on-site work is one of the key ongoing challenges in a GSA. Both budgets and knowledge levels must be considered. Budget constraints are crucial since tangible costs are reduced with increased offshore presence, even though the intangible costs of management overheads required for coordinating offshore work might be significantly higher. The transfer of knowledge about products, processes and practices to the offshore development team is also crucial since it gives the vendors a sense of trust that work can be done effectively at a distance. This onshore–offshore mix is not static, it shifts over time depending on the demands of either new work or peaks and troughs in the current workload. Firms try to deal with these variations through policies of staff rotation and the use of ICTs to maintain some form of 'presence' even in conditions of physical absence. As trust levels increase, higher-value activities may be undertaken but these activities in turn place demands for subtler forms of knowledge and understanding. We use the concept of *knowledge transfer* to refer to 'the transfer of knowledge to places and people where it is needed to be used to fulfil some activity or task' (Alavi and Leidner 2001). In our research, we rarely saw any explicit transfer of knowledge taking place directly from the offshore site to the client, other than the transmission of information required for project monitoring and control. The process of knowledge transfer in GSAs is extremely problematic and shaped by various socio-cultural–political dynamics, which in turn influence and are influenced by the broader trajectory of the GSA relationship.

The purpose of this chapter is to study the dynamics of knowledge transfer and the manner in which it interplays with the process of GSA evolution. The perspective is grounded in the ideas of a 'community of practice' (Brown and Duguid 1991; Lave and Wenger 1993). While knowledge is largely the property of individuals, some knowledge is regarded as being produced and held collectively and generated when people work together in tightly knit groups known as 'communities of practice'. Such communities are able to muster collective 'know how', the transfer of which is deeply rooted in practice, and through which members develop a shared understanding of *what* it does, of *how* to do it and *how* it relates to other communities of practice. The processes of developing the knowledge and the community are interdependent; the practice develops the understanding, which reciprocally changes the practice and extends the community. We use this 'communities of practice' idea to refer to the software development teams who are separated by time, space and cultural differences but still interact with each other using their common knowledge of various products, processes and practices. The community of practice idea is important to GSW, as software development is a knowledge-intensive activity that involves a large body of knowledge (*know what*) with a strong emphasis on practice (*know how*). Although the role of sharing knowledge and building trust has been described as being important in the functioning of communities of practice, little is yet understood of how distance, absence and cultural differences influence the effective functioning of these communities of practice (Little 2001) and with it the trajectory of the GSA process.

A 'community of practice' perspective on GSAs helps to go beyond a view that emphasizes the inherent capacity of ICTs to transfer all facets of knowledge required for software development. This 'information-processing' view conflates information with knowledge, as it assumes that knowledge is neutral, easily packaged, reduced and transmitted. Such a view is of limited usefulness in understanding GSW, since it presents a static and rationalistic impression of what in practice is a much more complex, socially negotiated and evolving process. To grasp this complexity, it is important to differentiate between information, learning and knowledge (Brown and Duguid 2000). Knowledge requires a person or *knower*; knowledge is harder to *detach* from the knower than information, as it requires *background understanding*. When an attempt is made to stretch communities of practices across time and space, it is this aspect of knowledge and the need for background understanding that make the process problematic. While the information-processing view remains a dominant one in information systems research, various critiques of it have been presented, arguing for the inherently subjective nature of knowledge (e.g. Baumard 1999), which has also been established empirically in different settings. For example, Lam (1997) describes a collaboration between a British and Japanese company and suggests that while one firm's design approach relied on experimentation, intensive interaction and learning by doing, the other used a formalized high-level design language. Although the first firm's design was not always consistent and logical, they had a better understanding of the tacit nature of knowledge than the second firm which used more formal methods.

Much of the knowledge in GSA is 'sticky' and difficult to transfer because of its *personal and tacit nature* (Polanyi 1962, 1966), thus posing limits on what can be effectively articulated and transferred. In a GSA, the knowledge to be transferred varies with the stage of the GSA process and is shaped by the use of different ICTs, which in turn raise new demands on knowledge. For Polanyi, while tacit knowledge is applicable in practical settings (like cooking), especially involving skilled tasks, it is equally applicable to mental activity such as language. Since formalizing and passing on such knowledge is problematic, it is best transferred in conditions of proximity where the learner can watch the expert, try to establish mentally what is done to make things work and engage in action through practice. Tacit knowledge (for example, recognizing a face) and explicit knowledge (for example, the height and weight of a person) are mutually dependent. Tacit knowledge forms the background necessary for assigning the structure to develop and interpret explicit knowledge.

A community of practice approach draws attention to the importance of tacit or background knowledge (Winograd and Flores 1987) and the need to integrate it with explicit knowledge, for effective knowledge transfer to take place between human *knowers* including individuals and groups. Examples of such integration are seen in Orr's (1990) study of Xerox technicians who formed tacit background knowledge through a process of face-to-face socialization, storytelling and 'hands-on' experience. This made the talk and the work, the communication and the practice inseparable. The Xerox technicians were co-located for at least some periods during the working week, making their learning deeply related to their everyday practice. There are, however, far different and greater complexities involved when the communities of practice are separated by time, space and culture. We study some of these dynamics relating to knowledge transfer in the case of Sierra and its Indian subsidiary and how these dynamics led to the shaping of the relationship.

7.2 Case narrative

Sierra is a small but highly profitable and rapidly expanding software house with offices in London (the headquarters), Brighton, New York, California and Bangalore. Sierra specialize in the development of short-life-cycle custom-built client server projects with customers and projects in a wide range of domain areas. In 1998, they had particular strengths in visual interface design and in 1999 moved into e-commerce development.

In 1998, Sierra employed 80 staff worldwide and company turnover revenue was approximately £8 million. Like many other specialist software houses, Sierra management were striving to overcome skills shortages and capacity problems in the UK. Their response was through a GSA arrangement in the form of a wholly owned subsidiary in Bangalore, India. Sierra believed that they could not pursue their strategy of corporate growth at their desired quality levels if they depended only on the development resources within the UK. The primary problem was the high salaries required for qualified staff.

Sierra's first move into India was the outsourcing of small projects to Indian software companies and involved some degree of 'body-shopping' Indian staff into the UK. A key problem in these early attempts was the large variability of quality because of the frequent movement of developers. Another problem was that the outsourced staff was not integrated into the Sierra culture – incentive schemes offered to Sierra staff were not offered to the Indian outsourced staff, for example. The differences were uncomfortable for actors in both organizations, and as time went on the development staff become factional.

Sierra felt that problems of inconsistent quality and cultural fragmentation could be addressed by opening up their own subsidiary in India. Following initial experiments with outsourcing, Sierra management was confident that the necessary skills to do software development were available in India, but to do it effectively there was a need to control the operations closely. A decision to open the India centre was taken very rapidly, based on a hastily constructed business plan. Within three months, a subsidiary was opened in Bangalore in 1998, with a modest growth plan from 10 to 24 people in 18 months. Early projects in Sierra varied in size and scope but all were short-life-cycle projects involving mainly data warehousing and interface design. We describe the process by which the GSA relationship between the UK headquarters and Indian subsidiary developed in two periods: a growth phase (1998–9) and a failure to stabilize phase (1999–2000).

Growth phase and attempts at stabilization (1998–1999)

The manager of the Indian subsidiary (whom we call Mitra) was of Indian parentage, born in Kenya and raised in England. He had a computer science education. Mitra and a colleague of his, who had been involved with Sierra's earlier outsourcing attempts in India, were responsible for setting up the Indian centre. These earlier experiences seemed to have provided Sierra with very little local contextual knowledge and no subtle understanding about 'the way things are done' in India. Mitra's view of the easy transfer of knowledge embedded in processes, and of the organizational culture, was important in shaping subsequent project events. Mitra adopted a broadly 'information-processing' view to knowledge transfer and believed that, armed with videoconferencing links, the Bangalore office could be made to mirror a 'little bit of Sierra' in India. Such a view reflected a gross underestimation of the complexities of bridging two diverse communities of practice in the UK and India.

The hastily conducted business plan justified the setting up of the centre on the superficial basis that the 'streets of India are paved with programmers', rather than on a deeper contextualized and realistic understanding of the complexities involved in attracting, retaining and motivating the programmers. Such insights would require an understanding of local nuances such as the nature of work practices of Indian developers. Maybe it was felt in Sierra that Mitra, with his Indian parentage, could bridge some of the knowledge gaps shaped by various contextualized processes. But Mitra had never really lived in India before and his frame of reference was very strongly UK-focused. The lack of understanding of local nuances and complexities led to great frustration in the early stages; for example, Sierra did not realize the need for a 'middleman' to negotiate with the Indian customs officials to release their videoconferencing equipment. The delays that resulted caused both shock and frustration and the starting of a realization that videoconferencing technology on its own would not be enough to connect India and the UK.

A significant issue in Sierra's initial plans was for the Indian team to take on whole-life-cycle projects including customer liaison. Unlike many outsourcing arrangements to India, Sierra's was innovative and willing to take risks by trying to establish a direct interface between the Indian developers and the end-clients in the UK. The belief was that distance could be made invisible under the assumption that the requirements could be easily encoded and transmitted using technology (such as videoconferencing, Netmeeting and email). As one person remarked, 'geography is history', and the Bangalore centre could be made to operate in a seamless way, appearing to their clients as Sierra UK. Usual procedures in the UK which involve regular face-to-face meetings with clients every 10 days or so would be replaced by a telephone connection through a leased line so that customers did not have to pay extra, which in turn would encourage communication. This model of communication was innovative as it was contrary to typical outsourcing relationships. These relationships usually start with small projects on-site at the customer's premises until sufficient knowledge is built up of local products and processes, and then some part of the development is moved offshore. Nevertheless the two expatriates felt that the knowledge gained through their early outsourcing experience, together with the ICTs in place, could substitute for the need of a face-to-face presence. They felt that knowledge could be effectively transferred. The knowledge issue was also compounded by the departure of Mitra's UK colleague who had worked on the Indian outsourced projects before 1998; he took with him some accumulated experience of doing work in India over the previous few years.

By the end of 1998, the Bangalore subsidiary had engaged in some full-life-cycle projects but even in those instances videoconferencing had not offered the levels of interaction hoped for. Technical reliability problems with the Indian systems – for example, frequent power failures – were a problem. Also, the only videoconferencing link was to the Sierra London office. This meant that Sierra's customers, who typically did not have access to videoconferencing equipment, had to travel some distance for the conferencing session. This was inconvenient as Sierra's customers were often located in other cities. The effective use of videoconferencing with customers thus required a great deal of logistical coordination of activities in terms of both time (difference) and space (multiple locations). The Sierra management had underestimated this need.

Communication difficulties also led to further problems in the transfer of information and knowledge. The UK customers and Sierra UK development staff found the 'strong' Indian accent difficult to understand. The difficulty was not helped by the unreliable videoconferencing and audioconferencing and emails which provided limited clarification and supporting gestures. The Indian team was sent on courses in the evening to 'correct' their accents, a process not liked by some. During an interview, one developer would switch between an 'English' and 'Indian' accent when speaking to the British and Indian researcher, respectively. When asked about this, he replied that he has two accents, one for working with the UK clients and one for home.

In one 'success story' related to us, a majority of the early life-cycle work (analysis and design) was done by an Indian developer who travelled to the UK client site for a few months to facilitate the process of eliciting requirements. The developer felt that such face-to-face contact was needed to reassure the customer and relate to the richness of the requirements. The initial interactions provided him with a shared frame of reference with the UK side, which helped to continue effective technology-mediated communication on his return to India. Maznevski and Chudoba (2000) have emphasized the importance of such initial face-to-face meetings in establishing a 'temporal heart beat' based on which subsequent meetings can proceed, even in conditions of temporal and spatial separation.

In addition to the problems of dealing with the time–space complexities in enabling effective knowledge transfer (see also chapter 6), there were further issues arising from the UK staff's perception of the Indian approach to problem solving. Owing to the dominant technical perspective, Mitra felt that the Indians ignored the equally important aspect of business requirements. The UK staff felt that the Indians overestimated their technical expertise, often reinterpreting the design requirements incorrectly, and sometimes unilaterally changing and removing things to deal with ambiguities. A UK developer said:

He [the Sierra India developer] didn't think it was important that things had to work in a very particular way. If there was a grey area, it was his job to make it not grey any more by implementing it in a certain way that unfortunately was not aligned with the original design.

These issues of 'incorrect reinterpretation' were magnified by the cultural embeddedness of certain assumptions concerning knowledge. Given its orientation of a small, global, high-tech company, Sierra UK had strong cultural norms about creativity reflected in their perceived ability to 'self-start' and 'think outside of the box'. The Sierra UK staff were expected to be able to use the software analysis and design techniques flexibly and think independently largely without supervision. They took pride in this ability. Mitra's and the UK-based senior managements' perspective was that such creativity was absent in the Indian developers who lacked a 'spark', and were unable to 'self-start' or formulate 'self-directed learning'. In the words of a senior UK manager:

It is a sweeping statement, but in India they don't tend to think outside of the box, they do what they have been told to do. And what that means is that if you miss something or something doesn't quite add up, they won't see it as a problem. They'll take it as read that that is what to do. So you have to be aware of that problem. I think that they will see there is a problem, but think that it is all sorted out. We want to encourage the developers to think outside of the box and to raise questions and issues.

Sierra's expectations, based on their previous experience in the UK and the USA, were that the Indian staff would be able and willing proactively to access and develop their knowledge of the technical products and user needs. Instead, the Indian staff was perceived as complying with methodology in a mechanistic, unquestioning manner. This was regarded as being an outcome of an emphasis on technical skills in the Indian staff as opposed to the lively, independent thinking, intensively questioning UK and US employees. In India, Mitra took particular measures in an attempt to remedy the problems by tightening recruitment with UK-style personal qualities overriding technical skills as a first requirement. In Mitra's view the Indian staff approach to problem solving might be questionable but Mitra's approach to evaluating creativity was also questionable. It was based on a 'Western' frame of reference. Mitra himself could be regarded as being uncreative, and not 'thinking out of the box' to understand the local norms of creativity.

The UK management believed Sierra India should mirror the non-hierarchical, free-thinking, informal and creative organizational culture of the other Sierra offices. Interviews conducted in the UK revealed the office to be laid out for 'hot-desking' and senior people would normally dress informally and share desks with junior staff. Mitra had high expectations of this informal, non-hierarchical structure being replicated in India and was very disappointed with the lack of success of his efforts to transpose the 'Sierra culture' and mirror this in the India operation. There were continuing problems with staff relationships and communication as Indian staff behaviour was seen to contrast starkly with the UK style of operation:

The London office is much freer. A typical example from the UK might be when someone joins us we open a bottle of champagne. People casually mill around, say hi and move on. Here in India I am expected to make a speech. 20 per cent drink, 80 per cent don't. We have some soft drinks for them to get people to loosen up. In the UK everyone drinks.

In a continuing effort to facilitate transfer of the Sierra UK organizational culture, staff from both locations were taken on a trip to the seaside resort of Goa in late 1998. However, this rather 'synthetic' attempt to create a relaxed atmosphere overall was difficult to translate into the reality of the office. Sustaining it in day-to-day office work was difficult because of the deeply culturally embedded nature of the work practices. Attempts at socialization between UK and Indian staff through informal talk (as in the case of Orr's Xerox technicians during their breakfast meetings) was limited in the ICT-mediated settings. The time difference between India and UK also meant that Indian staff felt excluded from the everyday activities of the UK staff. High staff attrition, a characteristic of the local context in Bangalore, also meant that a lot of energy was expended on constantly building new relationships and trying to repair the loss experienced in the knowledge base through staff departures. The UK staff working with Indian split teams indicated that communication problems were further compounded by the tendency of the Indian developers to avoid saying 'no', asking few questions and saying they understood when they might not have clearly done so.

The strategy of raising the technical expertise level by recruiting graduates from the more elite educational institutions also faltered because Sierra, being a small firm, could not hire a full-time HR manager. Mitra had to handle the personnel functions together with an Indian senior technical person. As a result, they could not invest adequate time actively going out to recruit from the major Indian university campuses where they faced competition for developers from the larger and more glamorous software houses of the MNCs. Sierra was thus left recruiting graduates from some of the second- and third-tier colleges where the primary focus is often on building technical skills rather than on broader education, emphasizing a 'technical' rather than 'business' kind of focus. Also, the skills of these graduates were rather outdated, with a majority of them having had their programming courses on COBOL, with little or no experience with newer languages like Java and Visual Basic. These outdated skills of graduates, coupled with their perceived inability to self-start and engage in self-directed learning, led the Sierra staff to believe they were spending far too much time on building the knowledge base of the Indian developers, which should have come as a given in the relationship.

The perceived continuing inability to transfer the Sierra UK 'relaxed' working environment had direct effects on the software development process. In Sierra UK, power relations were based on technical knowledge rather than hierarchical position in the company. It was considered legitimate for highly volatile 'creative discussion' to take place in meetings that were an integral part of problem solving. In these meetings a junior developer would feel they could openly contradict a senior member of staff. Mitra felt that in India the hierarchical position or 'tag' and related issues of status, title, rank and position were more important. Thus for the Indians hierarchy and position was relatively more closely related to knowledge: an older, senior manager's knowledge would have greater importance than that of a colleague or junior. Mitra's and the UK developer's view was that hierarchy was absolutely unimportant when problem solving: Mitra told us that he was uninterested in titles and rank and only in what 'someone brings to the table' in terms of direct contribution to solving the problem or issue at hand. This different status on knowledge had knock-down effects in the managing of 'creative meetings' both face-to-face and electronically mediated. Mitra repeatedly complained that in India he did not witness similar types of 'creativity,' and the nature of power relations was very different:

We are not allowed to enter into their [the Indians'] comfort zone. I am the manager: they will not let me talk to them, and they will always agree with me. In meetings the staff will often stay quiet and then I will get a lengthy email maybe an hour later when the meeting is over. And I have to start the thing all over again. And I think, well why not bring all that up in the meeting? It ends up taking twice the time. They will not confront me in meetings. The spark is missing.

The UK staff found it difficult to make the Indians explicitly reflect creative behaviour in the UK style. However, this form of creativity made little sense to the Indians. The reluctance of the Indian staff to follow UK practices of becoming informal with Mitra and to disagree followed by 'making-up' with after-work drinking, led Mitra to frustration:

It is really annoying. Sometimes I tell them something that is wrong. I want to hear 'I don't agree with you'. Here they are all interested in the 'tag'. I am more interested in what you bring to the table, not the tag you bring.

Another problematic area concerned the transfer of deeply embedded management processes that the Sierra management believed could be exported to India in the form of best practices, programmes and policies, including the language in which it was stated in the UK corporate Intranet. One form used for personnel appraisal in the UK contained the criterion of 'bollocking ability'. The term has a strong sense of meaning in the UK context – the ability to sternly discipline an employee when necessary – but had little meaning in India. While the Sierra management found this amusing in hindsight, the use of this term in the official personnel appraisal form was strongly indicative of a lack of sensitivity to contextual differences. Another example concerned the regular strategy meetings in London being filmed on videotape and played to the Indian team so they would understand the strategy process and plans for the organization. Indian staff found difficulty in understanding the accents, and were also deeply perplexed and confused by the UK team's supposedly volatile 'creative discussion,' which often had even senior people swearing and shouting at each other.

At a more macro level, new kinds of knowledge required for the e-commerce industry were placing pressure at the micro level of the India centre and the GSA. This need for heightened institutional reflexivity and Sierra's response to it is discussed in the next section.

Failure to stabilize and closure (1999–2000)

In early 1999 after several (mostly unsuccessful) attempts at whole-life-cycle projects being undertaken in India, it was felt that the model of work distribution should be moved to two discrete models, depending on the type of application. Model one, which had been in use prior to 1999 envisaged more interdependent work and used a split UK and Indian team. Teams were comprised of Indian and UK staff varying in size but usually small groups of 4–5 developers in the UK working with 2–3 Indian staff. Indian and UK staff either travelled back and forth or a local UK liaison offices was used to elicit requirements and deal with ongoing customer interaction. As well as the difficulties of communication related to the growth phase, lack of physical proximity was identified as a key problem with this model. In particular there were many difficulties relating to knowledge encoded into dialogue. Effective interaction with India needed instantaneous feedback and informal dialogue for informal, tacit knowledge transfer: Mitra described this need for interaction: 'When building software, a rapport is built up. Sitting next to each other passes on a lot of understanding.'

We develop this analysis further in later sections. In model two, the India team would assume responsibility for an entire module but the UK team would do most or all of the customer liaison. In this case, specifications would be sent to the Indian team and the code returned for testing and aggregation into the completed application. A mirror of the application as built was maintained at both sites. Videoconferencing, email and telephone as well as technologies such as Microsoft Netmeeting were used to facilitate communication between developers and, to a limited extent, customers. File Transfer Protocol (FTP) was used to deliver completed portions of code. Most work would be done with the latest World Wide Web application-building packages, JavaScript and database applications.

In model two, the strategy was one of replicating knowledge systems and minimizing interaction between the UK and India while model one emphasized replication and the development of knowledge transfer bridges through travel and the use of ICTs. Sierra management and development staff considered model two to be more effective owing to the modularization of work done in India, it could be self-contained and required relatively low interaction with anyone in the UK. This modularization and low level of interaction was seen as key to success but the replication of knowledge continued to cause problems.

Although Sierra management considered the strategy of replication of knowledge systems more successful, it was not without problems. India was perceived in the UK as having difficulty keeping up with the required pace and speed concerning time deadlines and the different manner in which these deadlines were interpreted. All of the Sierra group companies measured project management success based on efficiency reflected through 'leakage'. The UK management was constantly unhappy with the Indian developers who were seen to be 'leaking' on project deadlines. The Indian 'leakage' of 25 per cent was seen by the UK manager to be significantly higher than the UK office rate of less than 5 per cent. However, when the Indian developers were confronted with this concern, they were surprised that 'leakage' was an issue since they believed they had been delivering projects 'in time'. Further exploration showed that the main reason for the conflict was that the Indian and UK staff had very different time assumptions by which they measured 'leakage'. The UK office measured 'leakage' based on the number of hours worked on a project and not on whether or not a deadline was met. The Indians, on the other hand, measured 'leakage' based on whether they met a deadline or not. So it was quite typical for the Indians to work late in the evenings or over the weekend to make sure they met the deadline. However, the UK management still perceived this as 'leakage', as it was greater than the number of hours that were assigned for the project.

To analyse the reasons for these differences and problems in replicating knowledge systems more closely, we need to understand the different assumptions of time that shape social life in the different contexts. In Sierra UK generally there tends to be a relatively clear separation between home and work. A UK respondent said that when a developer comes in to work, he or she would not respond to a personal phone call or run out to do some domestic errand during work hours. The situation was different in Sierra India operating in a developing country context where work and home lives were more tightly integrated. Some developers had aged parents living at home and it would be quite common to take some time off in the middle of the workday to take a relative to the hospital. Or, since telephone bills cannot be paid by mail but only by physically visiting the office, it would be common for employees to take off some hours to pay the bill or to complete a transaction in the bank or post office. India has a relatively higher number of religious holidays and in the monsoon season it can be extremely difficult to travel. Further delays were caused by telecommunications failures. This 'lost time' was compensated by working late hours or in the weekends. Trying to apply the same criteria of 'leakage' based on assumptions of the UK work life context to India thus became problematic as each other's differences were not well understood or accepted. The result was that the India centre was consistently deemed as relatively less efficient than the UK, contributing to the lack of trust UK staff had of India operations and discouraging UK commitment to building a community of practice.

Further evidence of the problematic nature of replication of knowledge embedded in routines was experienced with respect to software development methodologies. Methodologies were in place but they amounted only to guidelines that were not rigidly adhered to as opposed to highly prescriptive structured methods. The Indian developers seemed happy with this freedom, but their interpretations of these routines were often to the letter and did not take account of UK cultural norms of permitted deviation from the standards. From the UK perspective, Indian staff was seen to use the guidelines in a mechanistic process rather than to question the validity of statements or stages.

In both models one and two, telecommunications delay and unreliability contributed to the reported difficulties in what was described to us as 'building a shared mental model' of customer requirements as the system was being built. One response under these conditions was to increase the level of formalism in design and communication processes and to have regular structured videoconference meetings. However, interviewees in both UK and India complained that there came a point when it became easier to do the job in the UK than specify it in the detail necessary to do it in India. A Sierra developer in the UK had the following opinion:

We have project status meetings and the only communication which takes place is asking things like 'Why are you slipping?' rather than discussing 'How are you getting on?', 'How are things?', a kind of conversation. When you do get through to India it is all 'What did you get to? 'What have you done this week?', 'What about next week?'. We miss out on things like 'I've just found this good bit of code which you might find useful.'

Informal 'corridor conversation' seemed important in transferring knowledge, especially for certain types of knowledge, and this was not possible in separation. A UK developer said:

We would have formal videoconference meetings with the India developers but a lot of the details would be worked out in an informal way between people. Only a small part of what you actually do gets written down and communicated through. This [informal knowledge] was important because we were closer to the customer.

Frequently Indian developers were seen to make incorrect assumptions about design documents. In some cases a UK developer found it quicker to complete the job himself rather than attempt to encode it or ask for clarifications. Misunderstandings were reported as being related to language, 'inferences' and 'base-level' understanding. Another UK developer said:

The communication, whether written or verbal, would be less rich because of language. The Indians speak good English, but you get the feeling that they are only getting say 70–80 percent and are missing out on any subtle points that might be inferred between people. If you describe something you often leave quite a few things unsaid that are assumed to be the case, whereas you have to be much more explicit with India. There is less of a base-level understanding.

By mid-1999, Sierra's business direction was being reoriented towards providing e-commerce solutions, as was becoming a norm in the industry, and being standardized through processes of globalization. The e-commerce area itself was so fuzzy that the clients themselves were not sure of what they wanted. As a result, Sierra saw the opportunity not only to implement solutions but also to define the problems. Sierra as a company consequently needed to pay greater attention to the business, to understand the ICT and information strategy needs of their clients, rather than just to provide technological expertise. This implied different knowledge requirements. There was subsequently a complex negotiation and rethinking of Sierra's strategic thrust. There was a general consensus that for the highly complex and early life-cycle projects which e-commerce applications entailed, the developers would need to be on-site with the customer. From the customer perspective, there were concerns about IP, especially in a new and uncertain area like e-commerce where ideas seemed to be valued more than the product itself. This was the primary concern which led Mitra to believe that Sierra customers would be fearful of outsourcing their e-commerce ideas and plans, as they might be stolen in remote outsourcing locations.

   To deal with the new strategic need for on-site work, and the ongoing problems of running the Bangalore office, Sierra decided to shut down its India operations and relocate the staff to the UK or US offices. Contributing to this decision was the fact that if the Bangalore centre continued, there would be difficulties in obtaining work permits for the Indians, and in trying to motivate the UK- or USA-based people who had worked on the first phase to come to India. Another personal and unexpressed reason was that Mitra by this time had become personally frustrated with the Indian situation and was keen to move back to the UK. Thus, at a macro level, a global shift towards e-commerce-type applications contributed to the need for the company to shift its operational emphasis. In Mitra's view this made the position for the Indian centre untenable as the early life-cycle work, conception, strategy and requirements for e-commerce applications required face-to-face presence.

7.3 Discussion

A number of themes relating to knowledge and its transfer were significant in shaping the trajectory of the GSA relationship. We draw on Blackler's (1995) 'images' of organizational knowledge which are useful to understand the different knowledge-related issues. These images include knowledge that is *embedded* into routines and systems or *encoded* into manuals, codes of practice and dialogue that are *embrained* and *embodied* within human actors and *encultured* in the norms and conventions of a culture. To analyse the case we draw on three of these images: encoded, encultured and embedded. *Encoded knowledge* is conveyed by signs and symbols and in GSAs they are realized in manuals, notations, standards and codes of practice, as well as information encoded and transmitted electronically using various ICTs. The debates around what constitutes *encultured knowledge* are complex but often simply defined as relating to 'the way things get done', more theoretically due to the process of achieving shared understanding. It is reflected more broadly in the structures, policies, norms, traditions, rituals and values of an organization (see for instance Smircich 1983 and our discussions of culture in chapters 5 and 9). While some aspects of this knowledge are formalized in rulebooks and codes of practice, other aspects are less formal and are shared during informal socialization and are socially constructed through ongoing and evolving processes of negotiation. There is considerable evidence of programmers and analysts approaching the IS development process differently in different countries based on varying encultured practices (e.g. Ein Dor, Segev and Orgad 1993). *Embedded knowledge* is that which resides in routines, technologies, roles, formal procedures and methods. In the Sierra case, development methodologies and project management plans, where knowledge is embedded into the routines, would be key examples. The following discussion will relate these images of knowledge to the case and analyse some of the dynamics of the initiation, growth and stabilization phases of Sierra's GSA.

Initiation

In the initiation phase, Sierra had outsourced several projects to India. Based on these experiences, Sierra management believed that they could do the outsourcing internally and decided that they could achieve greater levels of control over staff and the culture of the organization with their own subsidiary. They opened the centre based on a rapidly assembled business case in early 1998. Almost immediately, a lack of encultured knowledge of the Indian context, such as local formal and informal recruitment networks, bribes and corruption, infrastructure, educational institutions and bureaucracy, caused great difficulties. Informal knowledge about bribes, for instance, tends not to be formalized in official documentation about outsourcing, but is nevertheless an institutionalized practice in many Indian public and private organizations. Mitra did not discover this until he arrived in India; he then had to engage with the various systems that directly affected the process of software development. For instance, telephones took considerably longer to be installed and repaired than in the UK, bribes were required to get equipment through customs. Far from the 'streets of Bangalore being paved with programmers', as Mitra originally thought, he quickly found Sierra was a very small company in a large intense market competing for the best staff with large MNCs. Many companies would deal with this issue by hiring a local consultant. However, because of the strong desire of 'creating a little Sierra in India', Sierra started with a dominant UK frame of reference rather than a local one.

Sierra's early experiences with Indian outsourcing companies did not provide them with the necessary *encultured knowledge* to conduct business in India. This proved crucial to the lack of early success of the GSA. Owing to Mitra's lack of contextual knowledge, he created very high expectations from the UK side of what could be achieved in India; failure to meet these high levels led to discord and reduced trust from the staff in UK head office. Sierra initiated the GSA and attempted to transfer knowledge embedded into processes and products and believed that encultured knowledge could also be transferred by choosing the 'right' staff, benefiting from Mitra's influence and staging events like the holiday in Goa and evening drinks. Encultured knowledge was important at the macro level of the relationship as it provided background understanding and a framework for communication that could have been the basis for trust. Sierra management believed that their culture could be made explicit and manifest in the Bangalore centre.

The intention of taking control over the operations and culture through *standardization*, a theme also discussed in the Witech case, proved to be highly problematic. In the early stages, Sierra attempted to carry out a complex development offshore based on a strategy of standardization of knowledge systems. This strategy was quickly found to be wanting as software development work needed high levels of interaction with customers whose requirements were difficult or impossible to encode without face-to-face presence. When moving to the split-team model, physical separation still acted as a barrier to knowledge transfer between the Indian and UK teams. The UK development staff was closer to the customer and their informal interactions demonstrated the importance of the *situated* nature of development. This knowledge was impossible to transfer to India.

Growth and failure to stabilize

In the growth and failure to stabilize phase between 1999 and 2000, Sierra adopted three models of development. The first involved whole-life-cycle projects being taken by the India centre, including customer liaison armed with videoconferencing links.

This system was rapidly abandoned and development structured around a split-team model and an attempt at limiting interaction and replicating knowledge systems. Both models suffered from problems of knowledge transfer related to a number of technical and contextual factors.

Sierra's attempts to move to stabilization can be analysed from the perspective of their failure to create a community of practice, specifically with respect to issues of articulation of knowledge in practice or *knowing in action* and the importance of *common cultural background*. Brown and Duguid (2000) point out the importance of informal interactions, shortcuts and fixes to make the actors' form of talk and work mutually intelligible. Creating, learning, sharing and using knowledge appear almost indivisible. Such a dialogue did not emerge between Sierra UK and India. There was a constant need to check understanding that impeded spontaneous informal interaction. These problems were further accentuated by unreliable telecommunications. In a community of practice, knowledge travelling on the back of practice is shared. It is during practice, otherwise referred to as the act of *knowing in action*, where informal, tacit knowledge may be transferred. The Indians could not actively engage in such action owing to geographical separation, and the physical distance solidified the separate worlds of the developers situated in Bangalore and the UK. The UK developers who were closer to the client knew more than they could tell or specify formally in a reasonable manner short of completing the design in the UK. When they did complete the specifications they were often misunderstood.

Zuboff (1988) points out that knowledge encoded by decontextualized abstract symbols is inevitably highly selective in the representations it can convey. It is thus open to interpretation and can convey different meaning to different groups. At Sierra, poor interpretation of designs occurred owing to language and communication problems as well as differences in levels of shared technical and organizational knowledge. These differences were difficult for UK staff to detect. This led to some disastrously incorrect designs. For example, UK developers specified a workflow application and sent it via email for Indian programmers to develop. When they produced the specification they assumed that the Indian recipients would recognise basic workflow concepts and technology. For UK staff this knowledge was 'taken for granted', but Indian developers did not share this assumption. It took several costly iterations before the workflow application was eventually withdrawn from India and completed in the UK. The UK developers were under high time pressure and were unable to monitor the Indian developers' responses. They had no time to check understanding which at Sierra was a routine aspect of face-to-face interaction but difficult when mediated by ICT. If they tried, developers faced the time differential, the problems of making and sustaining a connection via phone or videoconference. Then there was the problem of telecommunications speech delay creating gaps in the 'conversation' and comprehension of accents. The problems were accentuated by a tendency on the Indian side to say they understood when they had not and discussion that lacked the informal tips, gossip about the customer needs and chat which was of such great importance in an evolving project.

In the Sierra case, the issue of encultured knowledge was important. It included an understanding of the levels of permitted deviation from methodology or other practices as well as the importance of time scales and quality, power and authority structures, levels of permitted informality, outside-work socialization, behaviour in meetings and in the working day. Cultural knowledge is related to history, habits, local context, role-related expectations and the prior disposition of individuals and groups (Tsoukas 1996). Like Polanyi, Tsoukas points out that all articulated knowledge is based on an *unarticulated background* of what is being taken for granted that is tacitly integrated by individuals. These particulars reside in the social practices and forms of life in which one happens to participate. For Tsoukas, it is when one lacks a common background that misunderstandings arise, in which case one is forced to articulate the background and explain it to others and to oneself. One knows the unarticulated background in which one dwells through having been socialized into it by others. This does not imply that individuals are all 'cultural dopes' caged by structural norms (Garfinkel 1967); although the individual retains agency, structural norms exist as tacit memory traces that are drawn upon in articulating action (Giddens 1984). The Gowing case described in chapter 8 also emphasizes the importance of background understanding. The Indian programmers working on social security systems had no conception of unemployment benefit or social housing. These were an unknown concept in the Indian system, but formed part of the taken-for-granted background knowledge in the UK.

The UK management style which celebrated informality, creative discussion and a relaxed atmosphere was difficult to transfer to India. At the same time the company was not able to foster a different kind of problem solving orientation for the Indians. The Indian practices were difficult to change in part because they were strongly rooted in the historical structures of the Indian technical education system. This tends to be relatively instructional: the focus is on learning techniques and methodologies. While this technical approach supports strong analytical thinking, it did not mesh with Sierra's 'out-of-the-box' approach to creativity.

Sierra management perceived informality, relaxed atmosphere and confrontation as being key to creativity. They found the Indians wanting in these aspects and were critical of the Indian approach to problem solving in the software development process. Sierra's creative conflict made little sense to Indian programmers; they tended to avoid open and direct conflict with the manager in meetings and preferred to send him an email raising issues after a meeting. By then, Mitra thought things had been resolved. Other writers have commented on stereotypical 'Indian' approaches to problem solving that emphasize group loyalty and cohesion, and trust and cooperation when trying to solve a complex problem (Badke-Schaub and Strohschneider 1998). The behaviour of the Indians was seen by Mitra to fit into the stereotype of Indians tending to be risk-avoiding and obedient, taking action only when some superior gives clear signals to go ahead. Mitra reported that the Indian programmers showed a greater reverence for hierarchy, rank and status: individuals were careful about expressing their opinions in a way that avoided disrespect for him as their manager. This tendency to value the positional role and hierarchy has been emphasized by other writers. Sinha (1984), for instance, comments that in Indian firms while obedience is expected from junior staff members, guidance is expected from senior leaders. Sinha and Sinha (1990) suggest that in many Indian firms there is a traditional differentiation of hierarchical relationships with the notion of 'check with the boss' as the crux of the decision making style. This shifts the locus of control into the highest position in the organization.

These issues of hierarchy had direct consequences for the development process and stalled the evolution of the GSA. Many aspects of the Sierra organizational culture had little if any connection with the background knowledge of the Indian programmers. Mitra's expectation that Indians would mirror the UK programmers was unrealistic. Initial face-to-face meetings are helpful to facilitate subsequent electronic mediation (Carmel 1999; Maznevski and Chudoba 2000). Achieving any position towards mutual understanding would require an appreciation of background and history of social practices and forms of life in which Indians and UK developers participate. Developing an appreciation of another culture requires extensive socialization, a genuine interest to learn and a strong investment of time and effort. For the UK side this would have meant an interest in Indian society, history, traditions and an attempt to understand Bangalore and India as a *place* rather than a *space* for software development, a theme discussed in chapter 6 in the MCI case. Sierra would have needed to adopt a process of *bi-directional knowledge sharing* rather than *uni-directional knowledge transfer* in order to develop a stronger encultured understanding. The Indian side would also need to attempt similar efforts to relate to the UK head office context, society, the customers, business, etc. Many large Indian outsourcing companies, such as Mastek, take this issue very seriously and run courses for customers and their staff so that they can aspire to a degree of mutual understanding.

The problem over 'leakage' shows the need to consider the broader context in which the Indian developers were working and how that context differed from UK conditions. 'Leakage' must be seen in the context of these differences and expectations must be realistic. Human geographers such as Massey (1995) point out the importance of place in the globalization debate. Some of the local contextual conditions of the place 'Bangalore' include the fact that it is subject to a strained, overcrowded infrastructure. During the monsoon season it can be extremely difficult to travel. Power cuts are common occurrences significantly affecting productivity. Differences in time zones between the UK and India and the relatively large number of vacation days in India also create issues in synchronizing team activities contributing to 'leakage'. Indian programmers often carry considerable family responsibilities that are quite different from workers in the UK, making the use of time quite different: late working at night or at weekends is quite routine for the Indians. Understanding such issues of working in India is fundamental to developing a *place-based sensitivity* to issues. Places contain existential significance for people living within them and form part of the tacit frames of reference that shape behaviour. For GSA to evolve, embedded knowledge needs to be viewed in the context of respective places and their structural conditions. In essence, companies like Sierra need to understand the culture of their foreign subsidiaries; their Indian employees need to understand the culture of their foreign clients.

Although the Sierra management faltered in their attempts to stabilize the GSA before finally deciding to close the centre, events show that knowledge transfer was not static. Various internal issues (e.g. communication problems of accents) and external events (e.g. a global and corporate shift to e-commerce) impacted and were impacted on by the knowledge transfer process over time. Communication problems led to a change in the nature of work conducted in the Bangalore centre that in turn reduced the need for communication. Corporate shifts to e-commerce caused a move to unstructured early life-cycle work that Sierra management realized could be done only in the UK.

We have adopted a broad sociologically grounded knowledge-based perspective to investigate the dynamics related to knowledge transfer. Three key areas have been identified to influence the process of growth of a GSA relationship – encultured practices, encoded dialogue and knowledge embedded into processes. Encultured knowledge of Sierra UK informality, creativity and conflict was seen to make little sense in the Indian context. The tacit background knowledge of programmers in the UK and India reflected structural norms in their respective countries. The backgrounds of Indian and UK developers were not sufficiently similar to allow for effective transfer. Problems of interpretation of encoded designs were in part due to an inability to relate to background tacit assumptions that could be made within situated (UK-based) groups. In table 7.1, we summarize these three issues and the nature of their influence on the evolution of the relationship.

7.4 Conclusion

Studying the process of a GSA from a knowledge perspective drawing on concepts of knowledge sharing in communities of practice provides rich insights into the GSA phenomenon. These insights help to develop theoretical and practical implications. Theoretically, there are insights into processes of globalization as perceived from a communities of practice perspective. The companies whom Sierra was attempting to emulate, such as Texas Instruments, Nortel, etc., are large ones with the capacity to make investments in the form of large numbers of expatriates, training programmes and other forms of standardization discussed in chapter 4. In contrast to these larger firms, Sierra represents the small and high-tech 'born global' firm described in chapter 1 as characteristic of the network society. These firms leverage the power of informational networks and intellectual capital to give them the potential to challenge the traditional large firms. Castells (2000) argues that such small firms still have handicaps of investments, personnel and brand name that they find difficult to overcome. Sierra, however, is an important contribution to Castells' argument. Their problem was not investment or brand image but *culture*. Networks on their own are not enough, and aspects of history, size and geography are still important factors with which they have to contend.

Table 7.1 Knowledge issues shaping GSA evolution

Issues of knowledge transfer	Initiation	Growth	Failure to stabilize
Encultured practices	Lack of informal knowledge of Bangalore context led to significant difficulties when engaging with systems and institutions (e.g. handling bribes, infrastructure, bureaucracy and recruitment networks) UK frame of reference for Sierra management	Sierra's practice of creativity, informality and behaviour in meetings made little sense in the Indian context where hierarchy, rank and status were important conventions Indian technical orientation was perceived as leading to a precise interpretation of methodology and unquestioning attitude to designs and specifications	Failure to transfer Sierra culture to India led to UK mistrust, dissonance and unwillingness to work in India centre
Encoded dialogue	Planned liaison directly with customers using ICTs failed owing to the location of UK facilities at Sierra London and the need for face-to-face presence to elicit requirements Telecommunication unreliability and accents caused problems of interpretation	Geographic separation led to misunderstandings owing to different unarticulated background of developers Indians could not engage fully in the act of knowing in a community through engagement in dialogue and problem solving ICTs were unable to bridge the Indian and UK communities of practice	Strategic moves to e-commerce required a shift to unstructured early-life-cycle work which could be done only in UK 'face-to-face' owing to the richness of interaction required Tacit background knowledge reflected different structural norms in India and UK which were not sufficiently similar to allow for effective transfer
Embedded knowledge	Standardization of knowledge systems in an unmodified form proved problematic owing to language and comprehension	Continued problems of standardization related to the Indian context 'Place'-related factors meant that replication of knowledge systems such as project management 'leakage' lacked relevance as a measure of success	The small size of Sierra made it difficult for standardization to be effective compared with large firms such as GlobTel

This is true for communities of practice where ICTs do not offer the complete solution to stretching these communities across time and space.

A knowledge perspective emphasizes the complexities and risks of GSAs, particularly to smaller companies like Sierra. It is naïve to equate knowledge with information and assume that difficulties can be overcome with ICTs. It is instead important to see knowledge sharing as involving '*human knowers*', rather than assume knowledge to be a commodity that is easily packaged, reduced, formalized and transferred. A human-knowers' perspective emphasizes the process by which knowledge is as a result transferred. The challenge for practitioners is to develop sensitivity to the importance of tacit knowledge and the difficulty of learning through practice when actors are separated. In GSAs, this means that ICTs must also support informal practices. It means a broader acceptance of the approach to heterogeneity within broader standardized templates.

Comparisons of the Sierra and GlobTel cases are interesting to illustrate some practical techniques and show how Sierra might have better managed the process. Sierra was a small firm, UK-based. Its relationship with the UK group was as a subsidiary as opposed to a JV. Sierra did not have the resources to move large numbers of expatriates to India to facilitate standardization in the same manner as GlobTel. Sierra did not have a 'big-company' reputation to draw the best graduates from the leading institutions such as IITs. Such graduates might potentially have had the kind of critical creativity that Sierra's approach required, towards the end of the centre's existence in Bangalore two such people were recruited who seemed to have the necessary skills and qualities of creativity and critical thinking. Sierra might have thought more about the role of 'straddlers' (Heeks *et al.* 2001) who would be a person or persons who could relate the interests of the respective communities to each other. Having sophisticated individuals who can bridge the cultures between groups, facilitate informal discussion and act as front-line liaison officers might have overcome some of the difficulties faced by Sierra and facilitated the practice-based learning experiences.

Sierra might have been better advised in their initial business planning to have considered alternative organizational relationships before hastily embarking on a subsidiary route. Entering into a JV arrangement with an established Indian company would have enabled them to set up operations with experienced people well versed in local and global practices and the limitations of GSW. This route, although with its own set of problems and difficulties, would have insulated Sierra from some of the knowledge transfer problems experienced.

A knowledge perspective also provides insight into the knowledge requirements for a GSA relationship to initiate, grow and evolve. Analysing the software development process in this way shows the importance of achieving *congruence* in the initiation stages. It demonstrates the need for an appreciation of tacit forms of knowledge, particularly in the realm of encultured practices which act as an interpretive framework for dialogue and for knowledge embedded in processes. As a result of continuing problems in this area such as lack of face-to-face dialogue and misinterpretation of designs, Sierra abandoned the idea of a workable community of practice between Indian and UK staff. Their view was that maturity in the relationship would have been achieved only by minimizing the need for interaction between UK and India by providing whole self-contained projects. However, evidence from other case studies points to flaws in Sierra's approach, especially with regard to their high expectations of Indian operations and the amount of work sent offshore. The first point can be related once again to Sierra's business planning, which did not provide a full analysis of the complex risks of working in India. UK staff had many preconceptions ('streets are paved with programmers') and high expectations ('little of Sierra in India'), which in practice were difficult to meet and had direct repercussions on management commitment to India operations.

With regard to levels of work done offshore, mature companies such as GlobTel set benchmarks for transition time of onshore staff to offshore, thereby managing knowledge acquisition and ensuring smooth transition to offshore. GlobTel also recognized the importance of encultured knowledge even as the relationship matured. They built in such mechanisms as reward and recognition systems aligned to relationship success, and tried to develop deeper understanding through frequent visits. In addition, overcoming differences in developer background knowledge may to some extent be achieved with regular visits, use of 'straddlers' or user representatives who may educate the offshore team on the context surrounding detailed specifications. These actions can possibly be reinforced by using such techniques as recording videos of the office context to be watched by the offshore team. The evidence from this case suggests that such techniques related to pre-planning, selection and training of people, effective communication and managing cultural differences are relevant to transferring knowledge. However the techniques discussed above are still not a complete substitute for situated community-based knowledge sharing through *engagement in practice*, which we argue still proves to be an important dimension of, and limiting factor in, GSW.

BIBLIOGRAPHY

Alavi, M. and Leidner, D. (2001). Knowledge management and knowledge management systems: conceptual foundations and research issues, *MIS Quarterly*, 25, 1, 107–33

Badke-Schaub, P. and Strohschneider, S. (1998). Complex problem solving in the cultural context, *Le Travail Humain*, 61, 1, 1–28

Baumard, P. (1999). *Tacit Knowledge in Organizations*, London: Corwin Press

Blackler, F. (1995). Knowledge, knowledge work and organizations: an overview and interpretation, *Organization Studies*. 16, 6, 1047–75

Brown, J. S. and Duguid, P. (1991). Organizational learning and communities of practice: toward a unified view of working, learning and innovation, *Organization Science* 40–57

(2000). *The Social Life of Information*, Cambridge, MA: Harvard Business School Press

Carmel, E. (1999). *Global Software Teams*, Englewood Cliffs, NJ: Prentice-Hall

Ein Dor, P., Segev, E. and Orgad, M. (1993). The effect of national culture on IS: implications for international information systems, *Journal of Global Information Management*, 1, 1, 33–44

Garfinkel, H. (1967). *Studies in Ethnomethodology*, Englewood Cliffs, NJ: Prentice-Hall

Giddens, A. (1984). *The Constitution of Society*, Cambridge: Polity Press

(1990). *The Consequences of Modernity*, Cambridge: Polity Press

Heeks, R., Krishna, S., Nicholson, B. and Sahay, S. (2001). 'Synching' or 'sinking': trajectories and strategies in gobal software outsourcing relationships, *IEEE Software*, 18, 2, 54–62

Lam, A. (1997). Embedded firms, embedded knowledge: problems of collaboration and knowledge transfer in global cooperative ventures, *Organization Studies* 18, 6, 973–96

Lave, J. and Wenger, E. (1993). *Situated Learning: Legitimate Peripheral Participation*, New York: Cambridge University Press

Little, S. (2001). Conclusion: managing knowledge in a global framework, in S. Little, P. Quintas and T. Ray (eds.), *Managing Knowledge. An Essential Reader*, London: Sage, 368–89

Massey, D. (1995). *Space, Place and Gender*, Cambridge: Polity Press

Maznevski, M. and Chudoba, K. (2000). Bridging space over time: global virtual team dynamics and effectiveness, *Organization Science*, 11, 5, 473–92

Nicholson, B. (1999). *The Process of Software Development across Time and Space: The Case of Outsourcing to India*, unpublished PhD thesis, Salford University, UK

Nicholson, B. and Sahay, S. (2001). The political and cultural implications of the globalisation of software development: case experience from UK and India, *Information and Organisation*, 11, 1, 25–44

Orr, J. (1990). Sharing knowledge, celebrating identity: community memory in a service culture, in D. Middleton and D. Edwards (eds.), *Collective Remembering*, London: Sage, 169–89

Polanyi, M. (1962). *Personal Knowledge: Towards a Post Critical Philosophy*, New York: Harper Torchbooks

(1966). *The Tacit Dimension*, New York: Anchor Day Books

Sinha, D. and Sinha, M. (1990). Dissonance in work culture in India, in A. D. Moddie (ed.), *The Concept of Work in Indian Society*, New Delhi: Manohar Publications, 206–19

Sinha, J. B. P. (1984). A model of effective leadership styles, *India International Studies of Man and Organizations*, 14, 86–98

Smircich, L. (1983). Concepts of culture and organizational analysis, *Administrative Science Quarterly*, 28, 3, 339–58

Telioglu, H. and Wagner, I. (1999). Software cultures, *Communications of the ACM*, 42, 12, 71–7

Tsoukas, H. (1996). The firm as a distributed knowledge system: a constructionist approach, *Strategic Management Journal*, 17, 11–25

Winograd, T. and Flores, F. (1987). *Understanding Computers and Cognition*, Bristol: Intellect Books

Zuboff, S. (1988). *In the Age of the Smart Machine: The Future of Work and Power*, New York: Basic Books

8 The case of Gowing and Eron GSA: power and control

8.1 GSW: a power and control perspective

In this chapter, we examine the role of power and control in shaping the process of a GSA relationship over time. In society and also in an organization, issues of power and control are intrinsically inter-connected with *culture*. Power has an influence on how cultural norms are collectively defined. The exercise of this power must be based on existing cultural values and assumptions. The exercise of power also enables the production and reproduction of cultural values. The process of globalization that defines the working of a GSA introduces new dimensions of power. Power structures that have been historically shaped – for example, through relationships between developed and developing countries – come into play in different ways and levels when firms from these countries are drawn together in a GSA. MNCs typically situated in the developed world have the economic and political power to make investments in infrastructure required for running GSAs. This economic power is also translated into cultural values, such as the norms of communication and conventions of meetings, etc. However, the introduction and stabilization of power- and culture-related values are always contested, especially in GSAs where the linkage between the firms is through the aspect of 'knowledge'.

Power and knowledge are also deeply inter-connected, a point made emphatically by Foucault (1991). Three key strands in Foucault's analysis relate to *power, knowledge* and *discourse*. Power is exercised through the discipline of individuals by the control of time (for example, time and motion studies), and space (for example, production lines) combined with standardization and surveillance of the drilled individuals. The second strand is knowledge; a society or an institution can be analysed through its 'regime of truth' or 'general politics' of truth that give rise to disciplines such as criminology. These disciplines create conventions, standards or norms. A deviation from these standards and norms is the basis on which one is categorized as a 'criminal' or 'pervert'. The third strand is discourse, where the first two categories are practically applied to the individual. Foucault's discussion of 'regimes of truth' is relevant to the analysis of ICTs that may be regarded as non-human political actors in the production and reproduction of knowledge, truth and power (Walsham 2001: 73). As with other expert systems (Giddens 1990), knowledge embedded into systems development methodologies contains regimes of truth; they metaphorically 'speak' on behalf of their designers – for example, the functionalist, structured methodologies such as SSADM (CCTA 1990) and interpretivist Soft Systems methodology (Checkland 1981).

Foucault's analysis may be linked to Giddens' conceptualization of 'expert systems' such as methodologies which potentially are instruments of power, surveillance and control, raising questions of which methodologies are used, who is imposing them and what possibilities exist to use alternative approaches. By opening the software development process up to surveillance of stages, steps and reporting in structured methods, software developers come under the all-seeing panopticon gaze of the managers monitoring the development process. In GSAs this is especially true, since structured methods are used widely as a strategy to deal with the complexities of separation by minutely controlling time, quality, project schedules, repetition in the form of the practice of programming, detailed hierarchies and continual analysis of deviation from 'normal'.

The *power–knowledge relationship is never static*, and varies with different stages of the GSA relationship. In the early stages of the relationship, the domain knowledge of what is to be developed rests primarily with the customer. As knowledge is transferred to the development team, the power differential is reconfigured and mutual dependency grows, the management of which is a key challenge in GSAs. The use of ICTs also enables other unintended consequences that can potentially reconfigure the power–knowledge relationship through a phenomenon described by Zuboff (1988) as 'informating'. The information generated through the use of ICTs provides greater visibility to the people and their actions, potentially making them more vulnerable to surveillance and control. Videoconferences help to monitor project progress, for example, but they also make visible various other aspects of the development process such as the messiness and chaos as developers try to manage the 'frontstage' in the meetings. This visibility can have unintended effects as the developers, feeling the pressure of being 'micro-managed', try to subvert these meetings. The power–knowledge relationship is also subject to *ongoing destabilization* through external events such as rapid technological changes in the industry or the opening up of new geographical markets. These changes place tremendous pressure on acquiring new knowledge. The groups having this knowledge hold power and this power furthers change as other groups such as colleagues and competitors develop similar knowledge.

Had Foucault experienced present-day globalization processes, he would certainly have been interested in the analysis of knowledge–power, discipline, surveillance and control across time and space. Managing such separation requires 'long-distance control' which Law (1986) has insightfully described in the case of the Portuguese navigators who made use of documents and devices and drilled people as a means of long-distance control in order to secure the global mobility of their vessels at sea and secure trading linkages:

Texts of all sorts, machines or other physical objects, and people, sometimes separately but more frequently in combination, these seem to be the obvious raw materials for the actor who seeks to control others at a distance. (1986: 255)

Standardized tables and methods of navigation, ships of appropriate build and the drilled 'model worker' described by Foucault as a 'reliable automaton', when taken together, offer a powerful way to exercise power and control across time and space. The use of ICTs such as software configuration management, contracts, service-level agreements, reporting mechanisms, penalty clauses and methodologies (for both software development and project management) similarly provide the potential for control in GSW. Giddens (1990) defines trust as a property of individuals and abstract systems or 'confidence in the reality of a person or system regarding a given set of outcomes or event[s]' (1990: 34). Other authors have discussed risk, trust and control and the implications for planning strategic alliances, IT outsourcing and virtual organizing (Faulkner 1999; Handy 1999; Sabherwal 1999). These authors in different ways describe how organizations strike a balance between the need for 'structural controls' in the form of written contracts and 'psychological contracts' that help to sustain trusting harmonious relationships.

The dialectical relationship between power, control and knowledge cannot be separated from issues of culture, a topic that is popular in both the academic and non-academic literature. The power–culture relation is reflected in the functionalist notion of 'managing culture' (Peters and Waterman 1982) shaped by the political intentions of corporate managers with the intention of exercising control (Child 1984). Culture is a 'slippery' notion treated both as something that can be 'managed' in a rationalistic sense and also as a 'spiritual' phenomenon. Smircich (1983) conceptualizes culture as something that an organization both 'has' and 'is'. Most often, the emphasis is on conceptualizing culture as something an organization *has*. This view may treat culture as a resource, like a 'creative workforce' or 'high-tech knowledge'. Culture is thus perceived as an objective entity which actors are able instrumentally to manage to the advantage for themselves or the organization. Such a functionalist perspective makes explicit the culture–power–knowledge linkage.

For our analysis, we were interested in examining how concepts of power and control were linked with culture, and all three together intertwined with different temporal phases of the GSA process. For this, we draw upon Giddens' (1984) structuration theory in which issues of power, culture and meaning are integrated in a subtle manner with an emphasis on the process of production and reproduction of social structures. Giddens analyses social life through three interconnected dimensions of *signification* (meaning), *legitimation* (morality) and *domination* (power). Social systems are constituted by the activities of human agents, enabled and constrained by the social–structural properties of these systems. These structures define both the rules guiding action and the resources empowering action, and exist only as remembered codes of conduct or memory traces. Giddens stresses that individual agents retain the ability to act according to will and responsibility (Whittington 1992). A key aspect of analysing institutions through structuration involves a focus on how structures come into being and not as artefacts or 'givens' of the organization culture. Culture is not analysed as something an organization *has* but as part of what an organization *is*, and the processes through which this is produced and reproduced. This draws attention to the personality of individual actors, the importance of managerial style in articulating structures of domination and the process by which culture is sustained in a continuing reproduction of relationships between individuals and groups of actors. Giddens recognizes a 'dialectic of control' whereby all agents will have some resources that they can use in a bi-directional process and thus even the seemingly powerless have some control.

It is only for analytical purposes that Giddens discusses the three structures independently. Conceptually he sees the domination structure to be linked with structures of legitimation and signification. Control restrictions are enforced in daily interaction, creating and confirming through a process of legitimation. In legitimation, the management of meaning (for example, values, mission statements) and the interplay between value standards of the culture and sectional interests of sub-groups reproduce and challenge the reproduction of structures. Norms and moral codes sanction particular behaviour and legitimize what is important and what is trivialized, thereby institutionalizing the reciprocal rights and responsibilities of social actors. Analysing who sets and reproduces legitimation structures and how they are articulated is important in an analysis of power. The signification structure is concerned with the cognitive means by which actors make sense of what others say and do. Thus language (signification structure) is drawn upon through cognitive schemes of syntax and semantics to create understanding; language itself is the outcome of these speech acts. Riley (1983) points to the institutional forms through which signification is organized as related to symbolic orders such as rites, rituals and customs which may be analysed in forms such as logos and architecture. Stories and legends (e.g. of hard work or humble beginnings) are also relevant to signification, as are slogans and acronyms. Analysis of everyday discourse displays metaphors and jokes that indicate the image members have of the institution and give clues to the inclusion and exclusion of social groups.

With respect to power, Giddens adopts a relational perspective viewing resources (allocative and authoritative) as facilities that agents are both guided by and draw upon in the exercise of power. *Allocative resources* arise from command over objects, goods and other material phenomena: for example, reward and motivation, budget and funds allocation and control of access to knowledge, information and technology. *Authoritative resources* are concerned with the coordination of the activities of social actors: organizational rules and regulations, policy making processes and methods and formal goals, objectives, and strategies, for example. Authoritative resources represent control processes involved in the structuration that restrict and maintain the reproduction of social systems. In GSAs, varying kinds of authoritative and allocative resources are manifest in the infrastructure that is established, the different methodologies and technologies in use and the various management control systems that are established to make the infrastructure, technologies and methodologies work in practice.

Control over these allocative and authoritative resources is linked to *knowledge and geographical domains*, a point that Castells (1996) has emphasized. Castells argues that in the network society, the 'power of flows' supersedes the 'flows of power'. This implies that power is no longer situated in traditionally significant institutions like the state and the church, but power is shaped by the networks an organization is situated in, the centrality of its position, its access to information and knowledge and the flexibility in its structure to apply this knowledge to innovate its internal processes. This conceptualization has important implications for GSAs in that there is the assumption that even smaller firms, powered by intellectual capital and their ability to leverage new ICTs, can potentially challenge the giant MNCs. However, as the Sierra case in chapter 7 pointed out, this is not always possible, as power and knowledge are still situated in structures that are *historically and socially specific*, and as such difficult to penetrate even in the network society. The relational perspective on power and culture in GSAs has to take into account cross-cultural issues because of the different groups, firms and countries that are involved. Robertson (1992) critiques Giddens (1990), pointing out that Giddens calls for an 'institutional' analysis as opposed to a cultural approach. According to Robertson, Giddens' focus on the institutional issues means that he largely neglects analysis of inter-state and transnational relations as well as international law and intercultural relations. Giddens' attempt to diminish cultural considerations is, according to Robertson, 'a great weakness':

While he may claim that globalization does not involve the crushing of non-western cultures he does not seem to realize that such a statement requires him to theorize the issue of 'other cultures'. (1992: 142)

Thus, for the purpose of this inquiry, additional writings dealing with cultural dimensions of globalization have been consulted and we have attempted to integrate cultural and cross-cultural issues into the theoretical framework. There is a large body of literature in international business that has attempted to theorize about these issues. These studies, many of which are based on statistical analysis of survey data, posit that distinctive differences exist between the occupants of a country because of a multiplicity of choices and priorities along cultural dimensions. Hofstede's (1980) popular study differentiated between various national cultural characteristics and represented them as dimensions. Analysis of culture and power can be seen to provide a blend in considering aspects of equality, status and individualism, for instance. This aspect is especially evident in Hofstede's study that categorizes national cultures along the dimensions of Power–Distance and Masculinity–Femininity. Hampden-Turner and Trompenaars (1993) studied twelve countries through the lens of seven oppositional pairs. They posit that culture and orientation to capitalism may be made up of a multiplicity of choices and priorities along these seven pairs. The analytical dimensions of *achieved status* versus *ascribed status* and *equality* versus *hierarchy* explicitly deal with attitudes to power across different countries.

We have not drawn on these studies directly because they have been subjected to extensive criticism on the basis of a standardized questionnaire, the assumption of a rationalistic perception of stereotypical cultural differences and that 'true' national characteristics can be derived from a stable, static milieu. Instead, to address cross-cultural issues within a structurational perspective, we draw upon Whittington's (1992) conceptual framework to link structure and agency. In this framework, managers are seen as being members of *multiple and often overlapping and conflicting social systems* from which they draw various rules and resources in the process of articulating agency. Such an approach to understand agency attempts to avoid stereotypes apparent in research that tries to define and reify 'national character' (Mead 1951) or the inherent cultural determinism implied in models such as Hofstede (1980). Even though structural conditions influence agency, all human beings are obviously neither the same nor 'cultural dopes' (Garfinkel 1967) unable to act outside of the caging effect of structures. Humans, through their reflexive and knowledgeable actions, are capable of changing these structures. Sahay and Walsham (1997) have applied Whittington's framework to the study of Geographical Information Systems managers in India, by conceptualizing managers as simultaneously being members of multiple systems including a community, a bureaucracy and a scientific community. In GSAs, this framework needs to be further extended to take into account the interaction between people from different firms and countries reflecting different political, educational, religious, community and class, familial and judicial structures. These structures shape managerial attitudes that are themselves influenced through the role of human agency. In GSAs, as actors work with others in multiple locations, they themselves are influenced and affected by contact with these global structures and over time the structures themselves may be reciprocally affected.

These are some of the issues we explore in the case study of Gowing and Eron. We use the related concepts of power and culture, drawing on a theoretical frame derived from structuration theory (Giddens 1984). We analyse how these power and culture issues allow us to interrogate the processes by which GSAs evolve.

8.2 Case narrative

Gowing Information Services (GIS, a pseudonym) are a software house and a part of the large Cass group of companies. Cass (also a pseudonym) are one of the top 250 companies in the UK *Financial Times* Stock Exchange companies with an aggressive, entrepreneurial style. Gowing was founded in 1995 as a series of software product acquisitions. In 1998, they employed approximately 100 employees and their turnover was around £10 million which made a contribution of around 10 per cent of the group profitability. Although part of the Cass group, Gowing are run independently.

Figure 8.1 shows the Gowing organization chart at the start of the research in 1998. Eron (also a pseudonym) is an Indian software outsourcing company with a 'software factory' located in Chennai. It was established in the early 1980s and is one of the top 30 software companies listed on the Indian Stock Exchange. Eron has a centre in the UK and in 1998 turnover was around £15 million across the group. It employed around 120 staff in the UK operations. Eron had grown largely as a result of the strategic initiatives of the visionary Indian directors of the firm and was serving markets in

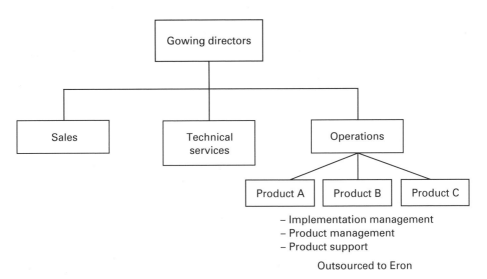

Figure 8.1 Gowing: organization chart

the UK, the USA and Japan. It had an ambitious expansion strategy to include other countries in its future marketing efforts.

While the Gowing management had no previous experience of GSA, some aspects of Cass' portfolio of businesses included outsourcing of various business functions. In that sense, Gowing's initiative in India was not totally without experience. The operations functions comprising implementation management, product management and product support functions were outsourced to Eron. In the early stages, this work was done with existing staff from corporate acquisitions but was eventually taken over completely by Eron and part of the work outsourced to India. Gowing serves the UK public sector with specialist accounting software that is based around relational database technology often linking into large mainframe computers for bulk processing. As will be discussed later, Gowing was formed from a series of corporate acquisitions and part of a group of companies. Cass perceived Gowing as part of a portfolio of companies it owned. The most important priority for Gowing, expressed by Cass directors, was making profits and shareholder satisfaction. Cass directors thus perceived Gowing as a 'money-making machine'.

Initiation (1995–1996)

In 1993, the Cass group embarked on an acquisition strategy and made an offer to acquire a specialist software business called RDC Ltd, as it supplied a market already served by Cass but not with software. The following year, David Jones joined Cass and subsequently acquired the rights to a product from PJ Computing Ltd for similar reasons. That same year Jones was made Managing Director of the new firm, now named

Gowing Information Services and remained until 1999 when he was promoted within Cass. Later in 1994, in a similar acquisitive move, Jones with backing from Cass acquired a small company, Hellenic Ltd, which had a product that complemented their existing portfolio. These acquisitions were important as they came with new products and some of the staff who had originally developed them as well, as in the case of RDC Ltd. Thus, the Gowing product portfolio consisted of three software products and the staff from PJ Ltd (shown as product A in figure 8.1, p. 161), RDC (product B), and Hellenic Ltd (product C) which were a result of three acquisitions making up the embryonic Gowing company.

The software products at Gowing comprised complementary accounting and financial reporting systems that were marketed to a range of large British public corporations. The corporations using these products had similar needs but sometimes customization was needed to take account of slight variations in the accounting conventions of regional offices. From its inception, the Gowing product portfolio was under constant change owing to revisions required by external bodies, involving legislative requirements on taxation, for instance. Other development work included maintenance operations and major updates, for example, from a menu-driven to a graphical user interface.

Gowing's office was located in RDC's buildings in a Devon seaside town. From the perspective of software-related staff, this was not a good location. Many highly skilled software staff were not willing to live and work in the area. They perceived the town to be in decline, inconveniently located, with few urban attractions such as cinemas, restaurants, etc. Interviews with several members of the staff indicated that their perception of the geographical area was a problem: 'There is very little to do here in the evenings and weekends: it is dead,' and: 'It is like dole-on-sea: there are many unemployed.'

Gowing is the only major software company in the local area and thus workers with relevant skills are not close at hand. The lack of any international population and associated networks also makes it difficult to attract foreigners. At the time of the GSA decision, there was a shortage of computer software skills generally in the UK, and it was becoming increasingly difficult to attract and retain staff. Existing staff was keen to move to city locations with higher salaries and better promotion prospects.

With this context in mind, David Jones, the Managing Director of Gowing, initiated the outsourcing of software development in November 1995. His stated motivation for outsourcing was primarily a resource issue, the desire to 'tap into' the large Indian software manpower pool. It was perceived that outsourcing to India could provide a logistical advantage in that people could be found at short notice as well as with a high level of English literacy and a large number of Indian computer science graduates could be hired at relatively lower costs than in the UK. According to Gowing's figures, UK programmers tended to cost on average 30 per cent more than Indian programmers. An important factor for Gowing management was the need to form an *organizational culture*, as the company was made up of a series of corporate acquisitions: the three companies had very different ways of operating. Gowing management perceived RDC as having had the strongest culture as it had been owned by the public sector and had what was perceived as a 'public sector' approach to software development. RDC's social relations and dress were informal and a 'kind of brotherhood' existed between the 25 ex-RDC actors. Work was consequently completed as a result of favour trading and goodwill, systems were often lacking in complete documentation and project planning was informal, overall documentation was considered of less importance than trust and completed releases of software.

The PJ product (A) was developed by a small youthful software house and was acquired along with the services of three contract programmers. Hellenic's product (C) also came with three staff. The staff from these two organizations were not used to extensive documentation of systems and detailed project planning. David Jones told us that: 'Staff who came with these different acquisitions had very different views of organizational life and of how software development should take place.'

Gowing management told us they wanted to form a homogeneous disciplined approach to software development work with an emphasis on efficiency and quality. This factor was another key motivation for outsourcing to India with Eron: it was perceived that Eron's emphasis on structured methodologies and disciplined quality management approach might help to tighten procedures at Gowing. A mutual friend of Jones and the Eron UK manager introduced the two companies and this led to initial meetings followed by a decision to 'body-shop' a small number of Eron staff into the UK. For Gowing, this move to body-shopping and involving the Indian programmers was a low-risk strategy to understand the possibilities of outsourcing and the extent to which Eron's disciplined approach could work in practice. Gowing management were optimistic at this point that a solution to their skills problems might be in sight.

The outsourcing activity began with four Indian Eron programmers on-site at Gowing's Devon offices. They were to assist with product development workload in product A (the ex-PJ Ltd computing product). Initially, this work involved mainly continued maintenance operations, which was a small project relative to the other products. Product A was seen as vulnerable as it was staffed by contract programmers who were expensive and could leave at any time. At Gowing, Eron staff introduced their quality methodology. It was a traditional 'life-cycle'-style methodology consistent with ISO 9000 accreditation requirements and prescriptions of structured, disciplined approaches to development and project management.

During this time, the Eron programmers were used in a 'body-shopping' role. They supplemented the work of UK staff involved in routine development and maintenance operations of the financial applications that were built on an Oracle database platform. Eron staff had to fully understand the application, their understanding was judged informally, at first based on their working on the development for a period of time and later by a written test and the application of set criteria.

Growth phase (1996–1998) and maturity (post-1998)

Over time, Gowing management gradually increased its trust in the competence and capability of Eron programmers. In this case trust was primarily gained by the application of an abstract system, a disciplined methodological approach to development as well as the adherence to procedures that the Gowing management perceived to be characteristic of the Indian Eron programmers. As a direct consequence of this, it was decided by Gowing management that a senior Indian Eron project manager would be moved from Chennai to Devon and the Eron staff would take over the Gowing product A development. Staff originally from PJ Ltd had drifted to other companies by this time and Eron Indian programmers were useful in filling the gap. After this, the model of work changed to an offshore model in which some staff and work were moved to Chennai. At this stage, roughly two-thirds of the Eron team (five staff members) were situated in India. Low-level specifications of work to be coded for the Oracle-based accounting product were sent by the Eron staff in Britain to the Chennai-based team who would return the code for testing. To facilitate communication and the passage of specifications and code, a leased line was in place between Eron's Chennai software factory and Gowing's offices in Britain. Email and telephone were used to clarify any issues or misunderstandings between the Indian team split between Britain and India.

As the Indian programmers had showed competence in the development of product A, it was decided by Gowing management that the Eron methods and employees were to be subsequently 'rolled into' Gowing product B which had been staffed by ex-RDC programmers. The implication of this was that the RDC staff would be required to use Eron quality methods under the supervision of the Eron project manager and comply with the standards imposed by the methodology. These standards were in the form of project management documentation of the adherence to deadlines and milestones. Other standards included rigid adherence to the detailed specifications, outputs and other documents of the various phases of the life-cycle-style methodology.

At this stage, the control objective of Gowing management was reflected in Jones' management style, described by some interviewees as 'ruthless' and 'mercenary'. His actions were legitimated by the Cass strategy of acquisition and profit making. The informal culture of RDC clashed with Eron's highly disciplined, transparent approach, ultimately leading to the demise of the RDC sub-culture. The ex-RDC programming staff ultimately resigned and there was a complete adoption of the Eron methodology and staff for all development work in Gowing. A period of intense development took place until the products were seen to be well understood and robust by all the staff. Eron staff now wholly develop all three products, both onshore and offshore.

Since the initial upheaval, leading to outsourcing and reorganization, the Eron–Gowing GSA relationship has become closer, moving from a 'body-shopping' model to one where considerable responsibility has been handed to Eron staff for all development work. This has included higher-value activities even at a strategy level as well as the full range of activities in systems development including requirements capture, design, development and maintenance. The two companies have intermeshed their activities, leading to a relatively mature arrangement. They have, however, continued to experiment with different levels of onshore–offshore development. These experiments involved attempts to move offshore more than the considered optimum of two-thirds; this caused problems as reducing the number of staff onshore in the helpdesk function led to intolerable pressure and workload on the onshore staff as well as problems of delays for fault rectification owing to the India–UK time difference.

Following the beginning of outsourcing, Eron was able to move up the 'trust curve' as it was perceived as performing well in the 'body-shopping' activities and as a result had been given responsibility for two major products. A new contractual relationship was initiated and the outsourcing project was moved fully into a stable situation that was still current at the end of the 2000 when we visited the company again. This implied a generally accepted maturity in the relationship.

8.3 Case analysis

At the inception of outsourcing, there was a conflict between the sub-cultures of the respective organizations that made up the embryonic Gowing company. The existence of these groups, in particular the RDC staff, was a major factor in the inception of outsourcing. The managing director of Gowing was keen to 'weld together' a coherent organization through what he called a 'sociological experiment':

Let me tell you about an interesting sociological experiment. I ended up with a mix of people, a bunch of RDC people, a bunch of ex-PJ people and some people from Hellenic. So I ended up with three different sets of people from different walks of life with different views of life.

These differing organizational realities stemmed from a fundamental difference in opinion about the nature of the software development process. The ex-RDC employees who had developed the original system represented the most significant sub-culture. Ex-RDC programmers, who represented a relatively informal culture, initially wholly staffed product B. Their informal dress style (sandals, open-necked shirts, in some cases long hair) clashed with Gowing management's desire for a formal style (business suits and tie). Relationships among RDC staff were less reliant on abstract systems, processes and documentation, and more on personal trust, kinship and tradition. These relationships were perceived negatively by Gowing management as existing in the form of 'a kind of brotherhood'. The ex-RDC developers viewed software development more as an art than an engineering activity, preferring to give less attention to the 'less creative' tasks of documentation and use of methodologies. They preferred 'hands-on' programming involvement instead, which they considered to be enough for the task. Used to a low-surveillance management style, RDC found that their approach clashed with the moves to a highly disciplined approach at Gowing. As Jones said:

In the RDC side of the organization, there was a certain arrogance about the way you design systems. And basically it was totally informal. There was no acknowledgement that project management, systems development methodology and standards were a good thing and so on.

The potential for the use of Indian programmers and structured development methods was seen by Gowing management as a way of facilitating change, improving control of development processes and aiding the creation of a 'corporate ethos' that could ensure the long-term survival of the organization. A key feature in this seemed to be the Gowing management perception that the Indian developers were 'more compliant, traditionally skilled, and less aggressive' than their British counterparts. This cultural feature was recognized by the Chennai team. According to the Eron Human Resources Officer from Chennai:

India is not a very assertive culture; Indians tend to go along with what other people say, especially with authority figures. When coupled with geographical separation, it becomes difficult, especially at Gowing.

Sahay and Walsham (1997) discuss how Indian managers and developers tend to be members of different social systems arising from both work-related and non-work-related systems such as intellectual groups, local community and family. Managers and developers in the process of creating agency and making action mutually intelligible, that in turn can potentially either reinforce or change social structures, draw on various rules and resources. Often, these rules and resources are conflicting; for example, the work norm of efficiency clashes with the family norm of helping a relative. There is thus constant tension and contradiction in the creation and articulation of agency. The caste system and norms of hierarchy often seen in Indian family relations are structural conditions that can be drawn on by Indians both implicitly and explicitly in developing agency. The caste system has contributed to value systems relating to status, power and relationships. Partly as a result, social relations are often seen to be hierarchical among Indians: people show status consciousness. In India, social relations exist between groups of a particular social standing. According to some writers, hierarchical structuring is so ingrained in India that it is often easier to work in a superior–subordinate role than as equals on contractual terms (Sinha 1988).

Bringing Indian staff into the UK context physically and 'virtually', because of the split teams, redefines and extends Sahay and Walsham's framework that was limited to structures within India. This situation of globalization presents a complex merging of social structures. For instance, a senior Indian project manager in his late 40s who had worked extensively in the UK and the USA managing software projects, while based in the UK, expressed the following opinion about Indians:

Most of our guys are submissive in attitude. They are shy. In India, you mingle exclusively with people of your own social standing. In the UK, that doesn't exist. Our programmers' behaviour changes when they come to Britain, they tend to be submissive even if they are authoritarian at home.

This theme of perceived submissiveness was repeatedly expressed, emphasizing the importance of incorporating norms of hierarchy into a structural analysis. The highly aggressive, competitive style of Gowing may also have contributed to the Indians' preference for accommodation rather than conflict and for their adoption of a more submissive posture. Many of the Indians working on the Gowing account were born into and follow principles of Hinduism that stress the virtues of contentment, absence of materialistic desire and stability. These teachings tend to oppose the dynamic striving for success and unlimited consumption that capitalist systems like that at Gowing emphasize. With regard to hierarchy, Roland (1984) states that in Indian work relations the superior is seen to be 'kind' and the subordinates 'submissive'. A consistent theme from many interviewees in this case was the Indian desire to please and to avoid confrontation. According to Gowing's Product Manager: 'When presented with a piece of work and asked if they can meet the deadline, the Indians will always say "yes", even when it can't be done.'

The above quotation indicates a general perception that the Indians feel a desire to please, especially in a situation of hierarchy that involves a sense of duty to the family and one's superiors. Sinha and Sinha (1990) make the point that in India, failure in one's role would bring shame not just on oneself but on the family as well. Our interpretation of the Indian developers was that their intentions were not simply ingratiating behaviour. More pragmatic reasons for the perceived submissive behaviour of the Indian developers relate to their supplier–customer relationship with Gowing. The structural controls in the form of the service-level agreement contained deliverables, reporting mechanisms and penalties for non-conformity with the contract. Eron staff lacked any significant trade union protection and the service agreement stated that Eron employees could be sent back to India if Gowing found them to be unsuitable in any way. In addition, the Eron team in Britain were working in a foreign land, dealing with accents that were in some cases alien to them, facing the uncertainty of immigration requirements and the absence of social and informal support groups which they would typically find at home in India.

The 'traditional skilling' of Indians that was of interest to Gowing reflects the emphasis on discipline in Indian schools with traditional drills, rote learning approaches and mathematical skills forming a large proportion of the curriculum. As a result, many Indians tend to be mathematically adept and disciplined in their thinking. The use of a structured information systems and project management methodology can be seen as power resources being used to control and create an organizational reality in accordance with the wishes of Gowing management. The Eron-structured systems development methodology formalized work arrangements at Gowing. According to Jones:

I could import a whole load of people who worked to methodology, project plans and were traditional in the way they worked. So they would sit there and wait for a product manager to bring a specification. If the product manager came over and chatted to them and then said 'Can you start now?' they would say 'No, not without a specification'. He would say, 'What's a specification?' They would say, 'You've got to write down what you want'. So he would go off.

The introduction of this structured methodology embodied by the Indian programmers helped to oust the RDC programmers from Gowing. The Gowing management perceived the RDC group as a counter-organization because of their relaxed, informal and undisciplined style of working. To quote Jones again:

Now we have got a rock of discipline right in the middle of the organization. So unless you've got the specification in here you won't get the code out here. Unless you put a test plan in, you won't get anything out.

Clearly the 'rock of discipline' referred to by Jones was imposed by the 'drilled' Indian programmers together with their documents and devices, including structured methods of analysis, design and project management. The use of the structured methodology when embodied by the Eron Indian developers contrasted with the established development style of the RDC staff. Jones' comments reveal his motivation:

The RDC staff didn't think we would, but we rolled the Eron methods into their product area. So that has meant that the rest of the organization has had to bend to the methodology and work this way. And that's how it works now. Even people who wouldn't do it in the past now subscribe to it.

The rigid structured approach encapsulated by structured methods of analysis and design is reminiscent of the 'mechanistic' organization portrayed by Morgan (1986). Software development and management were perceived by Gowing management as a machine-like process, where the programmers resembled replaceable machine parts. Importantly, the Eron staff at Gowing were trained to 'work to specification' which inevitably enabled greater control for Gowing management. This view was reinforced by metaphors in language used by Gowing management that included references to Tayloristic organizations such as Burger King as representing the pinnacle of a service organization.

Much has been written about the limitations of structured methods of analysis and design. However, their potential use as instruments of control, surveillance and coercion is of particular interest here. Winner (1977) defined technology as consisting of artefacts, techniques and methods of organization. He goes on to point out that a technology may contain assumptions and can be used to further the interests of those controlling it. Similarly, Latour (1996) in his Actor Network conceptualization, posits technology as a 'non-human' actor that carries the inscribed assumptions and interests of the technology developers and speaks 'on their behalf' in other situations. In this case, the methodology with its inscribed assumptions of structure and discipline is used by Gowing management as a resource of power to control organizational actors. Again according to Jones:

By getting this formal lump of formal process in the centre of the company, it has spun out. I can safely say now that everyone here subscribes to the idea of how systems are developed here.

It is worthwhile examining in more detail how structured methods were used in this case as a power resource to facilitate change. As discussed earlier, Eron brought the accredited quality methodology into Gowing. This is a structured development and project management approach that is similar to the traditional systems analysis life-cycle with products, deliverables and forms in keeping with many structured approaches. Wastell (1996) has argued that structured methodologies reflect a metaphor of the development process as a rational technical process embodying a rational engineering approach to system development. Baskerville, Travis and Truex (1992) also make the point that structured methods for IS development are based on scientific principles and thus embody a *reductionist paradigm*. In this way methodologies facilitate a control dimension by providing a coherent framework within which walk-through techniques, audit procedures, quality control, and inspection procedures can be incorporated (Ahituv, Hadass and Neumann 1984). This is especially significant as some of the Eron team were located in Chennai and were separated across time and space from their Eron colleagues based at Gowing's offices in Britain. The 'virtual' India-based part of the Eron team had no communication with the Gowing staff in Britain and their only contact was through their Eron colleagues. Such contact also meant task-focused specifications and the subsequent return of completed code; their virtual presence meant that they could not witness the impact or result of their work at Gowing in the UK.

When relating this analysis to Foucault's theorization, power was seen to be exercised through the discipline of individuals. There was a control of time using project management methodology, and a control of space by separating part of the Eron team which was in India and was not present to witness events. This control of time and space was combined with processes of standardization and surveillance of drilled individuals (Indian social structures that emphasize hierarchy) coupled with the use of routinized structured methodologies, which contributed to creating an agency that was interpreted by Gowing management as drilled submissiveness. The regime of truth manifested by structured methods is functionalist, and treats the development process as a reductionist, technical activity. Outsourcing development to Eron thus presented Gowing management with a new set of *allocative and authoritative resources* in the form of Indian developers. The Eron approach was seen as the 'best' way, legitimized the demise of the RDC developers and contributed to the wider utilization of the structured approach within Gowing.

The mechanization and division of labour brought about by the process was rejected by the counter-organization of the RDC programmers. This was in part because they perceived that the bureaucratic and rule-bound nature of the Eron approach helped to increase the control that Gowing management had over the development process. The standardized documentation helped to make the surveillance process transparent and visible. The transparency enabled Gowing management to determine whether the submissive 'drilled' Indian programmers were conforming to the prescribed procedures. Thus the methodology facilitated a division of labour across time and space and systematization of practices that allowed knowledge to be stored, systematized, disseminated and exchanged.

In an organization such as Gowing where the main 'product' is software, the Eron methodology helped to disassociate the Gowing management from the RDC methods of development described disapprovingly by Jones as the 'design on the back of a cigarette packet'. Instead they oriented development towards the Fordist production or assembly-line approach. Methodologies of this nature promote discipline among developers by specifying a structure for the development process and thus the transparency of work for surveillance. When coupled with the compliant nature of the Indian developers, the formal process of software production was laid open to Gowing management.

We now consider the context of globalization and how some of the power, control and cultural issues can be better understood within this backdrop. It is important first and foremost to consider some of the aspects of globalization that are reflected in this case. On the face of it, globalization provided Gowing with greater options in time and space to hire programmers to do their software development. As reflected in the quote below from Gowing's Product Manager, these global options were used by Gowing in a way that represents a *commodification of labour*: 'I don't care who does this work. It could be David Jones. It could be someone in India or it could be you – I simply don't care. I just want it done.'

At a deeper level, the course of events at Gowing provides an interesting example of the processes of globalization demonstrating some mutual and bi-directional effects. The effects of globalization are often discussed in terms of the impacts (quite often negative) that Western culture and management methods have in other cultures, a process that in recent years has been intensified by the effects of mass media. For example, the work of Ritzer (1995) identifies a burgeoning 'McDonaldization' of society, a thesis that essentially rests on a critique of the increasing pervasiveness of supposedly scientific, systematic and arguably dehumanizing management methods. Other authors (e.g. Beck 1992) have been critical of the exploitation of developing countries for cheaper labour costs owing to the ease of shifting production facilities overseas, and with it the 'redistribution of risks'. The implication of IT in the processes of globalization has led some authors to warn of the implications of 'electronic sweatshops' involving greater control and surveillance of workers (Attewell 1987). Others discuss the 'cultural imperialism' that rests on the assumption of cultural convergence created by ubiquitous Western mass media bringing images, symbols, products and entertainment into developing nations (e.g. Hall 1991; Martin 1995).

The Indian software analysts were trained in the rigorous ISO accredited methodology. The effect of globalization involved the disembedding of this Western-derived methodology into the Indian context where it was embodied by the Indian developers. The embodied methodology was then re-embedded into the British context by the Indian developers when they commenced work for Gowing. A similar example of the bi-directional effects of globalization is demonstrated by another case study of software outsourcing to India by Japanese firms in chapter 9. In this case, too, the Indian programmers, who typically have had prior software development experience with North American firms, were seen to bring in structured software development methodologies like the 'waterfall methodology' to the Japanese firms. The Japanese approaches to software development typically do not involve extensive documentation but rely more on discussions and personal face-to-face contact. The introduction of these methodologies is creating some disquiet within the Japanese firms; their managers are feeling rather uncomfortable with the changes that are being instigated in the move to written work. They describe the Indians to be 'too Westernized' and are contemplating changing the offshore outsourcing model to an on-site one where the need for written communication could be minimized.

A further effect of globalization manifested in this case concerns the two-thirds of the Eron team located in Chennai which was separated across time and space from their Eron colleagues based at Gowing's UK office. The India-based part of the Eron team had no communication with the Gowing staff in Britain and their only contact was through its Eron colleagues through task-focused specifications and the subsequent return of completed codes. The difference in time and space was significant because the India-based team was not made a party to any events taking place in Britain. Gowing management was presented with very different social structures, circumstances and global options than might have been experienced using a local outsourcing company based in Britain.

The general point being made through these examples is that GSW provides interesting examples of the nature of local–global interaction. Globalization effects are not necessarily going to be uni-directional from developed to developing countries, as in the traditional model, but increasingly *local events will shape global structures*. Future studies of GSW will need seriously to look at these local–global relationships and how they influence, and are also influenced by, the processes of GSA growth.

8.4 Conclusion and implications

All of the cases described in this book have been analysed to varying degrees from the power and culture perspective, as these concepts are related to most, if not all, human activities. We have illustrated this with some examples. As in the Sierra case discussed in chapter 7, management said that a major reason for opening the Indian subsidiary was to obtain greater levels of control over reward structures, motivation and creativity of staff. Subsequently, the issue of 'encultured knowledge' was a key factor in the failure of Sierra to maintain its offshore subsidiary. In the GlobTel case, the use of expatriates and training programs to standardize operations to the advantage of Globtel managers is another instance of how culture and power can be interlinked.

This case shows how a combination of the use of documents, devices and drilled humans using methods and information technologies for communication can serve as instruments of control. Compliant, and 'traditionally skilled', Indian programmers facilitated control both in Gowing UK and the Eron developers in UK and India. This is not the first study of power structures guiding and empowering action with regard to IS design and implementation (for example, Zuboff 1988). Markus (1983) was one of the early writers in the IS discipline to provide a comprehensive analysis of power within a longitudinal case study. The issue of power is also regarded as important in IS outsourcing, and there are studies that have addressed the issue. Lacity and Hirschheim (1993) and Kern and Silva (1998) analyse outsourcing cases focusing on power relations, but these studies do not explore global contexts, which is the contribution of this case. Previous related studies have also focused primarily on decisions of outsourcing and less on the management of GSA relationships over time.

A central implication concerns the need to understand the complex power, culture and control implications of GSA. Gowing still continues in its acquisition of new firms. In 1999, Jones was promoted within Cass to take control of several of the group's companies and has begun to implement the same GSA strategy. The GSA process has led to growth for the different organizational actors; Eron is taking up more sophisticated, strategic work, and is also now involved in other areas of the Cass group. The quality methodologies imposed by Eron have permitted a 'disciplined' approach to spread out to other aspects of Gowing and also Cass. As the contract has widened, Cass have exerted more financial control over Eron, enabling it to resist price rises and assignments of key Eron staff. The GSA relationship exists in a *dialectic of control* as Eron has built up significant knowledge of Gowing as its sole provider in an almost total outsourcing arrangement. Both companies are thus in a state of mutual dependency – Eron is dependent on the scale of the Cass contract and Gowing is dependent on Eron for internal staff to manage operations.

It is important to consider the implications of the analysis for the research and practice of GSA in India. The research implications include the provision of a framework for examining the issues involved in GSA that takes into account macro theory. Giddens' theoretical writings permit sophisticated analysis of the various structural forces that come into play as the process of software development becomes more globally distributed. This case study highlights the need to view this phenomenon in the light of a broader social context and against the backdrop of the effects of globalization. Specifically, the case analysis highlights the value of examining GSA from a *structurational viewpoint*. The case demonstrates the effects that globalization brings when sociological structures from different countries are brought together across time and space, and knowledge in the shape of methodology is disembedded and reembedded. The implications of globalization are seen to be both mutual and bi-directional. The case study also provides an interesting contribution to Foucault's analysis, drawing attention to the additional dimensions of time, space and absence with regard to the spatialization aspect of discipline and control.

In a period of extreme excitement and optimism regarding the so-called 'virtual' organization, this case study presents a contribution to that debate, sensitizing theorists to some of the unintended and also potentially negative and often overlooked implications. Walsham (1994) alluded to this point in an exchange with Mowshowitz (1994). Organization theorists or IS designers concerned with developing methodologies or frameworks should be made aware of the political and cultural implications of the outsourcing of knowledge work.

With regard to the implications for management practice, this case study provides insight into some 'information age' practices. The case explains how globalization provided the opportunity to circumvent the UK workforce and offered novel control possibilities to Gowing management. However, this strategy takes a view of organization that is concerned with the abstract requirements of a task and the required means which in a sense are stripped away from the task itself. Workers are seen in an instrumental way to be satisfiers within the process. This strategy represents a reinforcement of the Fordist methods of production that have significant implications for issues of long-term job satisfaction, respect for the worker and their loyalty to the company. This position has been extensively subject to critique elsewhere in the IS literature (e.g. Mumford 1983). The strategy at Gowing is particularly noteworthy, given that the most recent writings on the management of 'knowledge organizations' and 'network organization' call for partnership, trust, and cooperation with employees (e.g. Handy 1999). This case study offers a different picture based on a management paradigm focusing on narrow economic issues at the cost of the loyalty, emotions and aspirations of employees.

BIBLIOGRAPHY

Ahituv, N., Hadass, M. and Neumann, S. (1984). A flexible approach to information systems development, *MIS Quarterly*, 8, 69–78

Attewell, P. (1987). Big brother and the sweatshop: computer surveillance in the automated office, *Sociological Theory*, 6, 87–89

Baskerville, R., Travis, J. and Truex, D. (1992). Systems without method: the impact of new technologies on information systems development projects, in K. E. Kendall, K. Lyytinen and J. I. DeGross (eds.), *The Impact of Computer Supported Technologies on Information Systems Development*, Amsterdam: Elsevier Science, 241–69

Beck, U. (1992). *Risk Society: Towards a New Modernity*, London: Sage

Castells, M. (1996). *The Rise of the Network Society*, Oxford: Blackwell
(2000) Globalisation and identity in the network society: a rejoinder to Calhoum, Lyon and Touraine, *Prometheus*, 4, 109–223

CCTA (1990). *SSADM Version 4 Reference Manual*, Oxford: NCC Blackwell

Checkland, P. (1981). *Systems Thinking, Systems Practice*, Chichester: Wiley

Child, J. (1984). *Organization: A Guide to Problems and Practice*, London: Paul Chapman

Faulkner, D. (1999). Survey – mastering strategy: trust and control in strategic alliances, *Financial Times*, 29 November, http://globalarchive.ft.com/globalarchive/article.html

Foucault, M. (1991). *Discipline and Punish: The Birth of The Prison*, London: Penguin

Garfinkel, H. (1967). *Studies in Ethnomethodology*, Englewood Cliffs, NJ: Prentice-Hall

Giddens, A. (1984). *The Constitution of Society*, Cambridge: Polity Press
(1990). *The Consequences of Modernity*, Cambridge: Polity Press

Hall, S. (1991). The local and global: globalization and ethnicity, in A. D. King (ed.), *Culture, Globalization and the World System*, London: Macmillan, 19–40

Hampden-Turner, C. and Trompenaars, F. (1993). *The Seven Cultures of Capitalism*, New York: Doubleday

Handy, C. (1999). Trust and the virtual organisation, *Harvard Business Review*, 73, 3, 40–50

Hofstede, G. (1980). *Culture's Consequences: International Differences in Work Related Values*, Beverley Hills: Sage

Kern, T. and Silva, L. (1998). Mapping the areas of potential conflict in the management of information technology outsourcing, in W. R. J. Baets (ed.), *Proceedings of the 6th European Conference on Information Systems*, Euro-Arab Management School, Granada, 612–27

Lacity, M. and Hirschheim, R. (1993). *Information Systems Outsourcing: Myths, Metaphors and Realities*, Chichester: Wiley

Latour, B. (1996). *Aramis or the Love of Technology*, Cambridge, MA: Harvard University Press

Law, J. (1986). *A Sociology of Monsters: Essays on Power, Technology and Domination*, London: Routledge

Markus, M. L. (1983). Power, politics and MIS implementation, *Communications of the ACM*, 26, 6, 430–45

Martin, W. (1995). *The Global Information Society*, Aldershot: Gower

Mead, M. (1951). The study of national character, in D. Lerner and H. Lasswell (eds.), *The Policy Sciences*, Stanford: Stanford University Press, 70–85

Morgan, G. (1986). *Images of Organization*, London: Sage

Mowshowitz, A. (1994). Virtual organization: a vision of management in the information age, *The Information Society*, 10, 267–88

Mumford, E. (1983). *Designing Human Systems for New Technology: The ETHICS Method*, Manchester: Manchester Business School

Peters, T. and Waterman, R. (1982). *In Search of Excellence: Lessons From America's Best-Run Companies*, New York: Harper & Row

Riley, P. (1983). *A* structurationist account of political culture, *Administrative Science Quarterly*, 28, 414–37

Ritzer, G. (1995). *The McDonaldization of Society*, Thousand Oaks, CA: Sage

Robertson, R. (1992). *Globalization, Social Theory and Global Culture*, London: Sage

Roland, A. (1984). The self in India and America, in V. Kavolis (ed.), *Designs Of Selfhood*, New Jersey, NJ: Associated University Press, 170–91

Sabherwal, R. (1999). The role of trust in outsourced IS development projects, *Communications of the ACM*, 42, 2, 80–6

Sahay, S. and Walsham, G. (1997). Social structure and managerial agency in India, *Organization Studies*, 18, 3, 417–43

Sinha, D. (1988). Basic Indian values and dispositions in the context of national development, in D. Sinha and H. Kao (eds.), *Social Values and Development: Asian Perspectives*, New Delhi: Sage, 31–55

Sinha, J. B. P. and Sinha, D. (1990). Role of social values in Indian organizations, *International Journal of Psychology*, 25, 705–14

Smircich, L. (1983). Concepts of culture and organisational analysis, *Administrative Science Quarterly*, 28, 3, 339–58

Walsham, G. (1994). Virtual organisation: an alternative view, *The Information Society* 10, 289–92
(2001). *Making a World of Difference: IT in a Global Context*, Chichester: Wiley

Wastell, D. (1996). The fetish of technique: methodology as a social defence, *Information Systems Journal*, 6, 25–40

Whittington, R. (1992). Putting Giddens into action: social systems and managerial agency, *Journal of Management Studies*, 29, 6, 693–711

Winner, L. (1977). *Autonomous Technology: Technics-Out-of-Control as a Theme in Political Thought*, Cambridge, MA: MIT Press

Zuboff, S. (1988). *In the Age of the Smart Machine: The Future of Work and Power*, New York: Basic Books

9 Cross-cultural communication challenges: GSAs between Japanese and Indian firms

9.1 Background

The case studies discussed in this book thus far have concerned the work relationships of North American or European firms and Indian software companies. This chapter is different from the other case studies as it concerns relationships between firms from two Asian countries, Japan and India. The focus is also not on one particular relationship analysed over time. Instead a 'snap-shot view' is provided of a number of firms, in Japan and in India, that are engaged in or attempting to start GSAs. Since many of the relationships are not yet developed but in a stage of planning and projection, a large part of the analysis is based more on the managers' expectations of what issues will develop in the GSA rather than on actual experience. However, understanding these expectations is crucial because they shape the attitudes and actions of the people involved in the GSA. Although the cross-sectional research design is guided by pragmatic considerations of access and the intention of identifying and exploring interesting relationships and issues, it also reflects the current state of the business environment where GSAs between Japanese and Indian firms for software development are still in their infancy. While the success of East Asian firms in a variety of domains, including consumer electronics and automobiles, is well known, they are feeling the pressure to become globally competitive by strengthening the software component of their products. To respond to this pressure and to deal with the challenges of economic recession, these firms have started to build GSAs with software houses in India and China. They have also set up their own development centres in these countries (Sony, LG, and Samsung, for example, all have software centres in India).

In the last 3–4 years, some Indian firms have started to direct their attention to East Asia, especially Japan. While the Japanese software market is estimated at about $100 billion, the Indian share of it is only about $35 million. In 1998–9, only about 4 per cent of overall software exports from the Indian software industry went to Japan, as compared to 80 per cent from the USA and Europe (NASSCOM 2000). Through our interviews we informally gathered that in 2000 about 30-odd Indian software companies had established offices in Japan. This contrasts markedly with the high traffic towards North America involving close to 1,000 Indian firms since the early 1990s. There have so far been few sustained outsourcing projects from Japan to India, even though many new relationships seem to be on the cards. While a few rather tentative attempts have been made by major companies like Sony, NEC, Toshiba and Mitsubishi, these pale in comparison with the scale or long-term commitment of North American firms (GE and Cisco), or European companies (Phillips, Siemens, SAP, British AeroSpace, British Gas). However, this situation of GSAs may soon change owing to the general downturn of the American IT industry which has put pressure on Indian firms to adopt a strategy of 'not having all their eggs in one basket' and geographically redistributing the risks arising from an over-dependence on North America. Even the Indian software giant, Infosys, decided to reorganize their strategic business focus around geographical areas rather than technology domains, with plans to increase the Japan component to some 10 per cent of their overall business.

During the final stages of our empirical work with the North American and UK–Indian GSAs we became intrigued by the increasing discussion in the Indian firms, especially Comsoft, of the potential of developing alliances with Japanese firms. Managers and developers spoke positively and enthusiastically of the personal relationships that they had cultivated in the course of working with Japan, contrasted with the rather superficial and business-focused relationships with North America. India and Japan have similarities in religion and cultural values, such as respect for elders and the importance given to family. At first sight, it might be presumed that work relations between them would be easier to develop than with Western firms. Yet, as we began our investigations, we were struck by how thin the business relations actually were despite the inherent potential. To understand the underlying reasons for this, we conducted interviews in 2000–2002 with managers and developers from Japanese, Korean and Indian firms who were engaged in either Indian–Korean or Indian–Japanese GSAs. These interviews took place in the offices of these firms in Bangalore, Seoul, Singapore and Tokyo. During the course of these interviews, the respondents repeatedly emphasized that *cross-cultural communication issues* were the most significant challenge facing the GSA management. In this chapter, we attempt to understand these cross-cultural issues in more depth, focusing in particular on understanding their nature, why they occur, and the implications they have for the GSA relationship.

9.2 Culture, communication and GSAs

Researchers in international business have studied cross-cultural challenges to communication quite extensively, though mostly in face-to-face settings. Research has established the significance of communication issues in shaping the nature of business relations and practices. Hofstede's (1980) oft-cited study of cross-cultural differences between nations has also been drawn upon for analysing communication practices (Chesebro 1998). However, these cross-cultural issues take on a different form and level of complexity within the GSA context. The problems are magnified owing to the technical and knowledge-intensive nature of the language that is used and the multiplicity of products, technical processes, tools and methodologies involved. The use of different and often rather complex forms of ICTs, each with their own particularities and level of reliability, makes cross-cultural communication a significant challenge in the GSA context.

Cross-cultural communication research has generally drawn upon the concept of 'culture' in a rather simplistic manner, particularly related to the issue of unit of analysis. Researchers have primarily focused on comparing cultural systems across single or multiple dimensions (Chesebro 1998), using the nation state as the basis for analysis. Liebes (1988) empirically studied how people from different cultural systems perceive the same event (in her case, a TV show) and yet adopt different forms of narratives to describe it. She reported that while the Arabs adopted a more linear form for story telling, the Americans were more ironic and the Russians used a largely political perspective. Other authors have defended the use of the nation state as the analytical unit for pragmatic considerations of data collection (Chesebro 1998), or the need to generate 'stereotypes as hypotheses of national character' (Triandis 1972: 309), or to emphasize the search for universals and similarities, and not just differences, across cultural systems (Edelstein 1983). There is growing and widespread criticism of such research that equates culture with nation, on grounds that it is extremely functionalist and reductionist, and reinforces a 'nationalist–US-centric model of cultural exchange and interaction' (Ono 1998, p. 193). Similarly, Westrup *et al.* (2001) are critical of the Hofstede genre of research on the grounds that it promotes a static formulation of culture and ignores the processes by which cultures are constituted and maintained.

A functionalist perspective on national culture has also been a dominant orientation in IS research. Culture has commonly been treated from a social psychology perspective as something that differentiates one social group from another (for example, Schein 1984), or conceptualized as a variable that needs to be considered in the systems development process (Ein Dor, Segev and Orgad 1993; Shore and Venkatachalam 1995). Formulated in this way, culture becomes reduced to a set of variables that serve as a causal agent for explaining change. Recognizing such limitations, authors have in recent years argued for a view of culture that is contested, temporal and emergent (Avison and Myers 1995), and have focused on the processes through which culture is enacted, or 'accomplished' (Westrup *et al.* 2001). One conceptual approach to understanding such processes is developed in the Gowing case discussed in chapter 8, where cultural issues are addressed within a *structurational framework* that links social structure and agency. Culture is conceptualized in terms of the rules and resources that human actors interpretively associate with the various social systems (for example, family or organization) of which they are members. Human actors draw upon these rules and resources in the process of expressing agency, which in turn can alter (or reinforce) the actors' interpretations of these rules and resources. Examining the reciprocal linkage between culture and action in this way helps to focus on the process by which culture is continuously negotiated and 'achieved' in particular circumstances, rather than being assumed as a taken-for-granted given.

Communication, both face-to-face and ICT-mediated, is an important source of action in the context of a GSA relationship. When communication in GSAs is across cultures, various social structures, each with its own rules and resources, affect the process of expressing and interpreting messages. These messages can be in various forms – an email message, a telephone conversation, a videoconference meeting or a verbal or non-verbal message in a co-located setting. In choosing the appropriate medium and also in defining the content of the message, actors draw upon their understanding of the rules and resources in the various cross-cultural systems of which they are members. We may draw upon our understanding of whether we think it is appropriate to ring after 9 p.m. in a particular country before actually making the call, for example. Through the process of communication, actors both *reinforce and change* their interpretations of these rules and resources. Many of the Japan–India GSA relationships that we studied were in an exploratory stage; they had few prior established and shared frames of understanding to guide how communication should take place. In the absence of this prior experience of working and communicating together, we found the GSA staff often drew on their stereotyped understanding of the other to shape their communication processes. In this process of stereotyping, we found that perceptions of the national culture often superseded the organizational or more situated understanding of culture. In the Sierra case discussed in chapter 7, for example, the managers' image of Indians as being submissive to hierarchy was an important norm they drew upon in communicating with or about the Indians.

Although we understand and agree with the criticism made by academics of equating nation and culture and treating culture as some kind of a 'given' and 'objective fact', we argue that the tendency of managers to equate nation with culture cannot be ignored as the nation serves as an important source of rules and resources that they draw upon in the process of interacting and communicating. Taking these perceptions seriously is also emphasized in Anderson's (1983) argument about 'imagined communities', that nations exist only in *socially constructed imaginary spaces*. While this perceived or 'fictional' world is ontologically different from the world in which we live, it is significant in influencing our actions, including processes of communication. The quote below by a communication consultant advising firms in the GSA area reflects the importance of perceptions about national culture, used as a basis to develop communication strategies in GSAs:

As I am sure you are aware, different nationalities have their own inherent beliefs, attitudes, social systems and idiosyncrasies in the way that they operate and communicate. These differences can sometimes cause misunderstandings, and any company working in the international forum will encounter challenges in this area. (Shewell 2000: 7)

In GSAs, the structure of knowledge and the manner in which it is shared between group members is dominant in shaping communication processes. As argued in the Sierra case, the complexities of sharing knowledge about products, processes and practices is a significant and ongoing concern in a GSA. Based on a study of cross-cultural collaborations (between Japanese and UK firms) in the high-tech sector, Lam (1997) has argued that the *socially embedded nature of knowledge and organizational systems* can impede work and the effective transfer of knowledge across national boundaries. Lam's argument is based on the distinction between 'organizational' and 'professional' models of knowledge. Lam describes the organizational knowledge of the Japanese as being created internally in the firm through on-the-job and long-term training, similar to an apprenticeship where the focus is on an experience-based understanding. Such 'knowledge of experience' (Nonaka 1994) tends to be extremely tacit and context-bound, making it difficult to communicate easily to people outside the social and geographical context.

In contrast, Lam associates the professional model followed by the British with the existence of an external labour market where the focus is on acquiring general and standardized knowledge applicable to different contexts. The principles of learning are based on formal education in institutions, and tend to be more theoretical and abstract. The 'knowledge of rationality' (Nonaka 1994) is more standardized and explicit and thus easier to communicate in different contexts than in the case of organizational knowledge that tends to be more effective in face-to-face settings. The structure of organizational knowledge is more group-based, but the professional model emphasizes the individual. This distinction has implications for the coordination of work and the manner in which communication takes place to enable it. In the organization model, coordination takes place primarily through team-member interactions and is human network-based. In the professional model, however, knowledge is stored with individuals; there is strong reliance on documents and databases as a basis for communication. The discipline of software engineering can be seen to provide a guiding frame for such knowledge (Sharp, Robinson and Woodman 2000).

The organizational and professional models differ fundamentally in the degree of tacitness of the knowledge that needs to be communicated. Hall (1976) has previously attempted to relate the tacitness of knowledge with information required to convey meaning. Hall would argue that organizational knowledge is more tacit and context-based. It is, therefore, more difficult to communicate to members outside the social context than professional knowledge that is more explicit and context-free. Lam's empirical analysis drew upon the distinction between the organizational and professional models of knowledge to trace the asymmetries in knowledge sharing across the collaboration. Studying the processes by which knowledge is formed and organized within a community helps us to appreciate the asymmetries that exist and the challenges in communication to which they give rise. Inspired by the structurational framework described earlier, we differ from Lam, who treats the asymmetries as fixed and given, and instead conceptualize them as constantly changing and being renegotiated in and through the act of communication. Because of the cross-sectional rather than longitudinal design, we have not been able empirically to trace the mutual and changing linkages between knowledge structures and communication processes. The structuration perspective, however, sensitizes us to focus the analysis on the *processual and situated nature of communication*. We explore the processes of communication between the Japanese and Indian GSA members; we address the complexities that arise, how they are dealt with and how revised understandings develop. An analysis of this process can help to develop insights into the process of evolution of GSAs.

9.3 Indian–Japanese GSAs: some empirical issues of communication

Many of the Indian developers and managers we interviewed had previously worked with North American firms, and they often used this experience as a frame of reference to develop understanding in their Japanese relationship. In general, the Indians saw that the Japanese clients provided a greater potential for doing creative and higher value-added work earlier in the relationship than most North American firms. The Americans typically spent about 5–6 years on lower-value bug-fixing and maintenance projects before delegating work in core technologies. The Japanese were willing to assign projects involving new, sophisticated technologies on products that were at the critical stage of marketing testing early in the relationship. An Indian manager described the different approaches and motivations for outsourcing of the North Americans and Japanese as follows:

I think the Japanese . . . need and motivation for outsourcing is slightly different from the motivation of the North American countries that enter the market for reasons of cost effectiveness whereas the Japanese, to my understanding, is more a technology outsourcing. So, obviously, the relationship is at a much higher level as compared to the North American one where they already know everything about the product and you kind of pick up their knowledge. This puts them on a higher [plane].

The reference to 'power' is interesting as it indicates who has the potential to drive the communication process, and with it to enforce the rules and resources for communication. For example, in the GlobTel case (chapter 3), the Indians were on weaker ground because of their initial poor understanding of telecommunications. They had to depend on GlobTel for the knowledge and also the structure in which software development could take place. In new technology areas, however, there is more of a 'level playing field' in which activities can take place. For a number of Indian firms, the focus in Japan was on the high-technology niche sector of telecommunications where interesting and rapid developments were taking place in mobile and wireless applications. Innovative companies like Sasken (Bangalore) took a brave and futuristic decision to taper down some of their existing lines of work (mostly in the EDA sector) and focus primarily on telecommunications, with an emphasis on the Japanese market. Since Japan represents a key centre and testing ground for new telecommunication technologies and products, an increased demand for software services is expected to come from Japan in the future.

Our interviews, however, indicated that achieving this rich potential in Japan is extremely difficult. A key challenge both sides face is communication. We discuss three communicative challenges:

1 The structuring of the business model
2 The structuring of project management practices
3 The development of social relationships.

Communication issues and the structuring of the GSA business model

For both the Indians and Japanese, dealing with communication challenges was a key aspect in shaping the business model of the GSA relationship. Of course the primary concern was economic, but the business model was concerned with the need to deal with gaps in communication and to understand each other's practices. For the Indians who faced difficulties in understanding the Japanese user requirements, and for the Japanese in expressing them, an ongoing concern was how actively to develop mechanisms to minimize the 'communication and service gap'. Several alternative business models, some quite different from the Indian GSA relationships with Western firms, are being explored in the Japanese context. While it may still be too early to comment on the effectiveness of these models in practice, it is interesting to analyse some of their strengths and weaknesses in dealing with communication issues.

In our empirical work, we saw three main kinds of business models being adopted. The first was a *semi-customized product model* based on having semi-customized 'off-the-shelf' kind of product templates that the Indian firm can then customize with marginal effort to specific Japanese customer needs and products. The Indians examined these templates particularly in relation to the need for communication – minimizing the number of people who need to be sent for Japanese-language training, for example. Such a model can potentially help to reduce time to market and to increase value through enhancing the number of business transactions in the marketplace using the same product. In the telecom domain, protocol software such as Bluetooth needs to be adapted to new devices and platforms. It can be approached in a 'semi-customized product model' which does not require to go through the traditional cycle of understanding requirements and specifications and places a heavy burden on the need for intensive communication. The disadvantage of such a model is that the services market becomes downgraded – the market that has traditionally represented rather a lucrative domain for the Indians in GSAs with Western firms. Such a model typically involves a core group of Indian developers based in Japan with a working knowledge of Japanese; the on-site component of such a model would thus be small but relatively continuous with the primary responsibility being market development and providing customer support.

With the *domain-focused service model*, the Indian firm develops software for the Japanese company through the traditional process of understanding requirements, coding and development and finally testing and integration. The domain-focused service model runs into difficulties in business domains such as banking where client interface is important and constantly changing requirements place a heavy burden on communication. In this model, the Indian firm needs to develop human resources for the long run.

The Indian firm needs developers who are capable of speaking the Japanese language and have a reasonably sound understanding of the Japanese way of doing things. The CEO of an Indian firm in Japan that is relatively successful in using the service model described their level of investment in people:

We realized early on that in order to expand we have to familiarize ourselves with the Japanese way of doing business. We hired two professors from Japan and they ran a nine-month course for us in our development center. The engineers assigned for the course were housed there full-time along with the Japanese professors. Not only did they learn the Japanese language but they also learnt culture, dress codes, stories and music. It was a deep training. The software engineers are trained in both Japanese and technology. We have been quite successful. In addition, we have done a number of other things like hiring Japanese translators, hiring Japanese nationals at the front end, setting up an office in Tokyo, etc. We now have about 150 people who have gone through the training program.

In order to address the communication complexities that come with the service model, some Japanese firms attempted to develop a strategy of outsourcing to India only 'closer to the machine' (like compilers and operating systems) where development is largely guided by standard protocols and conventions, and interactions with users can be minimal. We quote from the interview notes made by one of the authors in Japan:

Three Japanese engineers attended the interview. Some extremely interesting points were highlighted relating to communication, approaches to software development, and the differences in approaches required for hardware development as contrasted to pure software development. Hardware-related development requires greater face-to-face contact as compared to pure software. First, hardware-related development is subject to much greater changes based on continuous feedback from end-customers. So, it was not possible to freeze the technical specifications before initiating the development phase of the project. Secondly, there is a much greater need to explain concepts using measuring instruments that cannot be done easily over the phone or email. So the next phase of work, that is going to be more hardware-based, the Japanese plan to do with a stronger component of on-site work rather than off-site. In off-site work, it was strongly felt that it would be only those projects that involved the use of standard specifications (for example, building C compilers or debuggers) where the contact points required for communications were minimal.

The above model will typically be operated with a small group of Indian developers, some of whom can speak Japanese, who will be onsite during the requirement analysis and also in the final testing and integration stage. Most of the development work is carried out offshore in India with the Japanese-speaking Indians trying to address the need for clarifications and questions by interacting electronically with the Japanese.

The *broker model* is one in which the Japanese company establishes a third party as a 'communication bridge' between themselves and the Indian companies. During the course of our research we saw brokers in different forms and sizes, ranging from the individual consultant to dotcom firms and even to larger conglomerates and subsidiaries. The individual consultants were typically retired Japanese officials who had lived in India and understood the working of the Indian system and made some key contacts in the upper ranks of some Indian firms. The dotcom brokers typically offered matching services of potential Japanese users and Indian providers registered on their database. Building business through this model was described by a dotcom operator as being quite difficult because of the importance of human contacts in Japan. An interesting new approach was the establishment of a joint venture called Japansoft (a pseudonym for a joint venture between a Japanese firm and three Indian software houses) to serve as a communication link between Japanese clients and three Indian software firms. An Indian engineer felt that Japansoft had placed a strong emphasis on communication, and although this is required, it downgrades to some extent the technical aspects of the project, leading to later problems in design and requirements analysis.

Another popularly used 'bridge' by Japanese firms is the subsidiary. Given the recession in Japan, firms have been obliged to support their subsidiaries to the maximum extent possible. A large Japanese conglomerate in which we took interviews were using their subsidiary in Singapore as the 'communication hub' through which to outsource software development to India. Managing this hub were Singaporeans of Chinese origin who were fluent in English. They could interface with the Indians and at the same time show some degree of understanding of Japanese communication. However, this model introduced a 'Chinese layer' between an already difficult to understand 'Japanese–Indian' interface. Many Japanese managers showed preference for establishing subsidiaries as it allowed them to communicate with people with a 'similar mindset'. We quote from interview notes made in a large Japanese firm that showed a preference to outsource to their subsidiaries:

The role of subsidiaries was emphasized in the context of outsourcing software development. The software engineers from the head office were described as sharing a certain common mindset with their subsidiary engineers because of their prior experience of working together. Used to this mindset, they found it quite difficult to work with the Indian engineers. For example, they were used to working on a discussion basis that they felt made it difficult for them to interact with the Indian engineers who seemed to prefer a more document-based kind of interaction.

The three models described above have different demands and needs for knowledge. Both Indians and Japanese need to adopt varying *organizational support structures*. While the service model can be seen as an attempt to deal with communication challenges by creating 'hybrids' who know the Indians and Japanese, the broker model attempts to keep the two groups independent and link them through a 'gateway' or a 'hub' arrangement. The semi-customized product model, in contrast, attempts to minimize communication interactions by focusing on products rather than services. In table 9.1, we summarize the associated communication strategies for the three models and their implications for managing the GSA.

The various business models adopted reflect the efforts of the Indians to access the Japanese organizational knowledge (Lam 1997) and of the Japanese to try and preserve it. The Japanese show a preference for the use of subsidiaries, as it helps them to deal with the internal labour market and people with whom they have worked before and who speak the same language. The Indians, in trying to access the complex Japanese organizational knowledge, try to revise their professional knowledge in two ways. (1) They redefine their

Table 9.1 Business models, cross-cultural communication and implications for managing the GSA

Business model	Communication strategy	Implications
Semi-customized product model	Minimize need for ongoing communication as implied in formal systems development methodologies	Reduce time to market by minimizing communicative transactions Downgrading on the lucrative service sector
Domain-focused service model	Support traditional software development methodologies with extensive training and education on Japanese language and culture	Japanese prefer 'closer to the machine' projects Extensive investment in learning language and culture
Broker model	Use of a 'person' or organizational entity as a communication 'bridge' or 'hub' supposed to understand the language and working systems on both sides	A 'bridge' introduces its own complexities Technical aspects seen to be downgraded in quest for making communication strong

professional knowledge relating to systems development methodologies by adopting a 'semi-customized focus' that does not require the traditional development cycle. (2) They supplement their professional knowledge by making heavy investments in language and cultural education, much more than they would when working with North American firms. These models make varying demands on the Indians to understand the Japanese culture and way of doing things, and also on the Japanese to develop expertise in formal systems development methodologies. While the service model places maximum demands on the Indians to learn Japanese, there is equal pressure on the Japanese to understand structured systems development approaches.

Communication issues and the structuring of project management practices

Communication issues vary with different aspects of the project and include the processes of initiation, requirements analysis and reporting control systems. The manner in which these issues are handled has significant implications for the functioning of the project and also the GSA. We discuss these issues related to three different facets of project management:

- Project initiation
- Requirements analysis
- Project operations.

Project initiation

One of the firms we studied described the extremely detailed and long-drawn out process before the Japanese decided to start a GSA. Hampden-Turner and Trompenaars (1993) described this approach to relation-building as 'polycularity' representing a 'gradual and seemingly circuitous route in which the Japanese build their relationships' (1993: 114). A senior manager felt that negotiations for a contract with the Japanese executives are more difficult and time-consuming than with North Americans. Contracts from North Americans are like a 'low-hanging fruit'. The large teams of Japanese company executives who come to visit and negotiate are prepared with detailed background information about the Indian firm that is often unknown even to the Indian executives from the same firm. Japanese executives are seen to bargain hard and negotiate to the smallest detail in project costs and schedules. An Indian executive described in amazement that 'they even specify Tokyo or Bangalore time'. The same executive described the negotiation process to be very demanding, with the Japanese even employing specific 'wearing-down' tactics like extending a meeting for 14 hours. All aspects of the project need to be negotiated down to a fine level of detail, but once finalized, subsequent changes are strongly resisted by the Japanese. In contrast, the Indian developers with working experience in North America described the American clients as relatively more relaxed, less formal and open to variations in estimates and project deviations as long as they were informed in advance. An Indian female manager described that, while the Japanese were meticulous in deciding to enter into a GSA relationship, the situation changed quite radically once the initial trust and confidence had been established:

The trust was built maybe in 3 years, that's a lot of time. First they say we are not having many projects, and we will keep you open. But once that happens, they just say, 'I want this project and can you do it?' It doesn't go through the formal processes with the marketing department making presentations. They will just directly pop an email to one of us and say, 'Can you do it, and how much is it going to cost me?'

Requirements analysis

Central to the task of requirements analysis is the process of communication and how effectively firms manage them. The Indians described the interactions during the requirements stage as being extremely time-consuming and taking place mostly in co-located settings. This face-to-face interaction was considered necessary to establish the basis for further communications in the project, which occur mostly over email, rarely by telephone and almost never by videoconference. The specific approach for communication adopted for the requirements analysis varied with a number of issues, including the kind of project, prior experience of working together and the platforms on which development needed to take place. One project leader who had worked with different clients said that it was more crucial to do analysis when they were working with firms that did not have standard development platforms. As many of the Japanese did not, available documentation was normally minimal (and in Japanese), thus emphasizing the need for analysis, and placing extensive demands on translation and communication. However, in other cases where there were more standard platforms in place and there was also prior understanding of working together, an Indian engineer said: 'We don't really have to submit our specs to them. They just give it to you and we have to implement it. That is all they have given us. That is [how it is].' However, in general, many of the Indians echoed the feeling that they tried to keep the communication to a minimum:

So they always prefer the minimum requirement we come out with. So they prefer less interaction. That is one thing that leads to a lot of confusion because we perceive something and then we deliver it. Then they see it and say this is not what we wanted. We have come out better when we have face-to-face meetings. We try to always do that, however time-consuming it is.

In general, the Indians described the interactions with the Japanese to be extremely time-consuming because of the problem of language and the extremely detailed manner in which the Japanese checked requirements. An Indian engineer described his understanding of the language problem in the following way:

The key issue out here I see are the cultural differences and different ways of doing things. There are language barriers. Japanese are perfectly okay with the written language; we give them the documents or mail. They will read and send it, be able to respond, but in face-to-face meetings it is not so.

This quotation reflects a rather superficial understanding of the Indian engineer, as he is not able to tie up the language issue with a deeper cultural understanding and a broader analysis of the Japanese way of doing things. For example, another Japanese manager describing his reasons for preferring the Chinese to the Indians for GSAs said: 'The Indians may understand some part of the spoken language, but the Chinese also understand the written and *unwritten* language.' The emphasis on the unwritten aspect reflects the importance of the cultural context and its intertwining with expressed and unexpressed language.

Project operations

After the requirements stage, which normally takes place at the Japanese site, the Indian engineers typically take the development offshore to India and work from there, while trying to maintain minimum electronic interaction with the Japanese clients. Sometimes the Indian engineers need to go to Japan to deliver something or obtain clarification. A senior Indian executive of a large Indian firm who had had experience with a range of offshore clients said that their company had an onsite component of around 5 per cent in Japanese projects which contrasted quite sharply with the 20–25 per cent figure with North American firms. While the number of interactions was few, when they took place they were extremely detailed and time-consuming. In frustration, an Indian engineer said:

They go through in detail. They take much time. I won't be surprised if they go through the code even. They go line by line through the program they write. We were so surprised when we sent our engineer; they had four days specifically for code reading. And they went through it line by line.

Despite the emphasis on detail and the time that interactions took, the Indians did not see the reporting and control systems of the Japanese to be as obsessive as those of their

North American clients. A senior manager who had worked previously with the GlobTel account said: 'We have project reports and reviews, but they are not as tight or religious as in the case of GlobTel. The method is more like, "We have given you the jobs and what we expect on the day of the delivery".' In general, the Japanese system of communication over the course of the project was described as oriented more to outputs rather than to the process milestones which are typically specified by the software development method-ologies. The focus on output, while giving the developers a sense of freedom in contrast to the complaints of 'micro-management' against the North Americans, also led to prob-lems in some cases. The Indians felt that in the final stages of project delivery the Japanese would tell them that this was not what they had asked for. While the late detection of such problems could potentially have been minimized with closer control of the process, developers often associate closer control with the fear of a loss of freedom. Also, some of the rather young (and often more brash) Indian engineers attributed the Japanese lack of emphasis on process to be a sign of their failure to understand software development.

In general, the Indians approached the project using their professional knowledge of formal system development methodologies, and various quality certification standards (ISO and CMM) to which their firms conformed. This professional knowledge had over the years been reinforced and developed in working with the Americans who had emphasized a similar structured approach to software development. This adherence to professional knowledge could be seen as the reason why the Indian engineers in their interactions with the Japanese would emphasize written documentation, including project plans, specification documents and project deliverables with well-specified start and finish dates. In contrast, the Japanese would typically start without a formal plan or a specification document; many of them expressed in interviews a preference for business based on discussion rather than formal written communication.

This preference could be understood in relation to the historical experience of the 'employment for life' system where employees shared the same mindset about how things were done, and requirements need not be written out, since a person could always walk across to a colleague and seek clarification. The Indians' constant emphasis on a document-based communication put unwanted pressure on the Japanese engineers who were largely unused to working in a culture of reading and writing long documents. Historically, the Japanese have shown a seeming indifference to formal contracts, relying more on the 'spirit of the law' rather than the 'letter of the law' (Hampden-Turner and Trompenaars 1993). The Indians, on the other hand, were more guided by the 'letter of the law' expressed through formal methods. The Indians were often working with Japanese hardware engineers who seemed to have a greater indifference to structured 'waterfall'-type methodologies than the software engineers.

In interviews in both India and Japan we were told that the Japanese showed a preference for pictures as compared to text. Hampden-Turner and Trompenaars (1993) write that the 'very existence of the comic books for the educated shows Japan's greater reliance on whole pictures and images rather than printed words alone' (1993: 123). We observed this preference in interviews where the Japanese respondents would often reply to a question by drawing a picture and we also saw young and old people alike engrossed in reading picture-based comics in subways. This preference for graphics and texts was at odds with the Indian style of writing long text-based memos and reports. An Indian engineer remarked:

One of the greatest learning [curves] we have is how to express themselves as much as possible graphically rather than in words. Describe something if we can try [it] out and explain it to them, that way you get the message across. The biggest [problem] we have out here is we are dealing with a customer or with a partner who, because of the very nature of the work, is not able to give very clear specification.

Indian software managers felt that although the Japanese in general conform to procedures and work processes defined in detail, they appear to face problems in following software work processes. The Indian programmers seem to be more proficient at this, probably as a result of their stronger base of 'professional knowledge'. Consider the following two quotes from an Indian software manager:

The Japanese are very proficient in the hardware business but not so in software. The Indians are more systematic, they have procedures in place. The Japanese do not follow [steps of software processes] strictly. They know what the functional design, etc. is but they do not follow it. In software, they do not have so many procedures.

Highly structured and relatively slow-to-change processes are well executed by the Japanese. Software processes, however, need to be fleshed around a skeleton of rigid process. This needs flexibility and initiative that is difficult in the traditional Japanese approach.

In general, the Indians felt that highly structured and stable processes were well executed by the Japanese. Large Japanese computer companies such as Hitachi, Toshiba, NEC, Fujitsu, etc., attempted to organize software work along lines similar to manufacturing activities. The resulting so-called 'Japanese software factories' that were initially successful (Cusumano 1991) did not survive rapidly developing technology. As some Indian managers commented: 'The Japanese are very strong in legacy systems but they are very badly positioned in terms of recent skills. They are in crisis.' A possible interpretation of this crisis could be the relatively slow and limited uptake of the ICTs in the course of GSA work. A common assumption made in GSAs is that ICTs normalize problems of communication across cultures. Walsham (2001) has challenged such claims, arguing that homogenization is difficult, if not impossible, given the extreme diversity across the globe. From the Indian perspective, we found their use of ICTs to be guided by two interconnected considerations: (1) understanding the Japanese way of doing things; (2) dealing with the issue of understanding and speaking the Japanese language.

The 'Japanese way of doing things', which can be interpreted on the basis of the 'organization knowledge' model, is extremely difficult to understand, especially for the young, technically qualified Indian engineers who strongly believe in the supremacy of their professional knowledge of software development, a view reinforced by their relative success with North American clients. One Indian engineer complained about the Japanese 'group approach' to software development – when there is a bug in the software, the entire team is responsible for finding and fixing it, rather than just the person responsible for the particular module, as would have been the case in a North American project. This 'group responsibility' of the Japanese working style also means that multiple rather than specialized skill development is encouraged (Hampden-Turner and Trompenaars 1993). The young Indian engineers, used to the emphasis on specialization and clearly demarcated modularized work, found understanding the Japanese people to be much more complex than understanding the North Americans:

One thing about the Japanese is that it is very difficult to read. And the Japanese people, I just don't know what there is in their mind. Is very easy to read an American. A Japanese will be so polite to us but we don't know what there is in his mind. It is difficult to know who controls in Japan and who makes the policy decision . . . in North America you know who really calls the shots. Here it is not that open. It tends to be more collective, and it does not look like there is one person. You would not know with whom to campaign, as you don't know very clearly where the buttons are.

In trying to deal with the complexity and apparent ambiguity of culture and decision making, young Indian engineers tried to communicate minimally and very directly, and related only to technical issues, by specifying keywords and leaving management issues to someone more experienced. A young engineer with 2 years' experience with Japan described communication processes during the course of a project:

I communicate only on technical issues. I am not involved in the management level. We try to simplify the language as much as we can, and with every mail we write that 'If it is not clear, you can come back to us'. Videoconferencing, too, we use only for technical reasons.

Another engineer emphasized how long it took to communicate even relatively simple issues. They had to repeat each question two or three times before it could be understood. As a result, he said: 'I do not use the phone much, and given the option I would . . . use . . . only email.' The manner in which the email text is structured from both sides is quite different from the experience of communicating with North Americans. The Indians try to write very brief, simple, short sentences where each line only makes one point. And while the Japanese reply often takes longer, it is meticulous in dealing with all the points raised. An Indian manager remarked:

If you ask them a question, it is not that they won't answer as very often is the case in the USA under a similar situation. You ask five questions in a mail, one would fall in the crack. When you ask five questions in Japan all five question will be answered meticulously, even if it takes 2–3 weeks to do it. Because the mail would probably be sent to somebody who can translate English to Japanese and then the Japanese engineer will ask him to translate from Japanese to English. The translator can always overload it.

Along with the long time taken in communicating, there are also misunderstandings, ranging from the harmless and the funny to the more serious. One Indian manager jocularly described how a Japanese manager told him, 'You are welcome to Japan and we will hospitalize you'. An Indian lady manager gave examples of more serious misunderstandings:

A bottleneck was communication, and generally we have misunderstandings. The requirements text was given to us about an order in which we were to design a chip, and they always said, 'that we had to down the data, down the data'. And we perceived it as though the data were right. And this was happening for almost 7–8 months. But actually what they meant was 'read the data'. And we were sending it thinking that it was all the correct answers and we did not know why. Then we visited them and realized the misunderstanding. Another example was they would say, 'it has to support all the layers'. While for us 'all the layers' meant what is there now, for them it meant also all the layers that it can have in future...

The Indians seemed to rely primarily on intensive initial face-to-face communication and subsequently on email, which they tried to keep to a minimum. Often the requirements were sent to them by hard copy or by fax rather than email. Phone calls were used but not extensively. Even though Indians may have learned some spoken Japanese, what is spoken over the phone is very different. In fact, we were told that private institutions run special language-learning classes for 'telephone Japanese'. So, the Japanese, who are used to speaking on phone primarily to Japanese people, told us that they had great trouble in shifting to the mode of talking to Indians who have learned Japanese in classes primarily intended for face-to-face interaction. Videoconferences were rarely used because the voice was deemed to be not as clear as on telephone, perhaps because of bandwidth constraints.

The choice of using a particular ICT is tied to other issues, including how the Indians perceive the Japanese way of doing things and vice versa. These perceptions significantly shape the choice in the use of technology ('never use videoconferencing') and also the mode of expression ('write short sentences with one idea'). These choices have broader implications for the project – for example, on the percentage of offshore and on-site work – which is in turn related to the question of attrition, as described by an Indian manager:

In GlobTel we were working with the North Americans, and thus you get a chance to go to the USA and work there. That one thing is more attractive and lucrative for the engineers than Japan. But here we do not get much chance to go abroad once the development goes on in India. There are certain cases when people have gone to Japan but only for short periods like a month or two, while with North American clients we can go for a year or two, make some money and come back. That is a critical fact that makes one decide whether or not to work for a North American client.

The choice of technology is not simply of whether it is used or not. The choice is intricately tied up with a range of *socio-cultural issues*, some of which have already been discussed. In table 9.2, we summarize the key communication challenges at various stages of the project, and how individuals tried to deal with them.

Communication issues and the development of social relationships

Communication and social relationships cannot be separated from each other, and the analysis of communication is fundamental to understanding social structure (Couch 1989, 1996). The manner in which we communicate – written or unwritten – and

Table 9.2 Communication challenges in project management

Project phases	Communication challenges	Response to the challenges	Role of ICTs
Project initiation	The process leading to the decision to start a GSA is very detailed, long-drawn-out, time-consuming and exhausting	Large groups of Japanese visit India Long and detailed meetings Indians try to revise their expectations based on the North American experience of picking 'low-hanging fruit'	Minimal as the primary reliance on building a relationship is based on human networks
Requirements analysis	Problem of expressing requirements in English Indians' insistence on detailed, frequent written documents resented by the Japanese Japanese show preference for graphics to text Indians specify start and finish while for the Japanese these distinctions are more ambiguous and blurred Non-standard development problems with minimum documentation (or in Japanese)	Initial meetings in co-located settings Indians set up a core team, few of whom may speak Japanese Indians try to develop as detailed requirements as possible to avoid electronic contact later Some Japanese seek 'closer to the machine' applications to try and minimize communication	Minimal in the initial stages as face-to-face meetings preferred Once trust is established, projects can be started without detailed requirements specifications Email used minimally to clarify requirements Sometimes documents, especially in Japanese, sent by fax
Project operation	Japanese focus on outputs rather than processes often causes problems at the time of project delivery Indians' need for documentation resisted by the Japanese Excessive time taken to answer queries because of translation issues Frequent misunderstanding in meaning	Frequent project reviews but not as 'religious' as North Americans Indians changing strategy about their use of ICTs, both medium used and content and structure of message Indians revise expectations of how much and how email should be used	Email is preferred medium, phones less and videoconference even more limited Drastic redefining of structure and content of messages – brief, simple, short, less frequent

interpret and respond to communication is fundamental to shaping our feelings of *cohesion and reciprocity* with others. In GSAs, the situation is complex, as a significant aspect of the relationship and interaction needs to be managed in conditions of a 'virtual presence', where members, especially when work is being done on-site, need to share consciousness of each other through some combination of textual, auditory and visual contact. The degree of reciprocity in communication, what Couch (1989) describes as 'social responsiveness', reflects an understanding of the mutuality of interests, beliefs and emotions, without which 'there is a minimal merger of self and the other'. The extent of this merger has implications for identity which is created and experienced through the 'negotiation and co-construction' over 'meanings and manner' among members interacting in a specific context (Wynn and Katz 1997). Identity, when congruent or shared, reflects the sense of 'oneness' among the actors in a collective, irrespective of their personal biographies or geographical locations (Couch 1989). A shared identity permits organizational actors to perceive themselves as part of a larger and more autonomous whole. Formation and maintenance of such a shared identity is a prerequisite for future oriented cooperative action, especially where there is a high level of task interdependence among the teams (Knoll and Jarvenpaa 1998).

Social relationships thus include a number of aspects, including a sense of cohesion, a feeling of co-constructed identity and mutuality of interests and emotions. While our empirical basis is too thin to substantively comment on these complex issues, we make some tentative observations about the issue of communication and social relationships. In the Indian–Japan GSAs, written and verbal communications are seen to be extremely complex and time-consuming, especially compared to interactions with the North Americans. There are challenges in speaking the language, understanding meanings and the amount of time and effort required for even 'simple' transactions.

Other difficulties in building social relationships arose from broader differences in working styles and, also interestingly, eating habits. The Japanese tended to work extremely long hours and nearly every day, in contrast to the Indians, who worked fewer hours and had frequent holidays. This was often interpreted by the Japanese as meaning that the Indians were too casual and not careful about adhering to deadlines. The frequent turnover of the Indians meant that the Japanese had to interact with a new individual that made it difficult to develop enduring social relationships. Indian engineers, frequently vegetarians, were often reluctant to travel to Japan for extended periods because of the difficulties in finding vegetarian food and also the exorbitant cost of dining out. Often, the Indian engineers would prefer to sit at a different dining table to their hosts because of the radically different food preferences. In Japan, where social relationships are often best developed over food and drink, when sitting together after work, the Japanese managers often felt that they had limited opportunities for the team members to bond together. Another issue was age; typically the Indian engineer would be younger, fresh out of college, as compared to the Japanese manager. The

Japanese culture emphasizes respect for elders, and the wishes of elders are accepted unquestioningly. The younger Indians felt that such structures limited the expression of their creative energy. They felt that these structures made acceptance and openness to learning from past mistakes difficult. The Japanese interpreted such attitudes as the Indians being 'too Westernized'.

Despite the challenges of language and limited physical and electronic contact, it was very interesting to note that the Indians expressed very positive feelings about the social relationships and understanding they had with the Japanese. A key reason for this seemed to be the underlying sense of a *personal* touch in the enduring communication that extended temporally and socially beyond the official boundaries of projects. This made the relationship much deeper as compared to the Indians' North American relationships. The 'personal touch' is created because the Japanese do not separate the official and formal from the unofficial and informal; this gave the Indians a feeling of personal involvement and an engagement that transcended the work relationship, which to them was very significant. A sense of trust was built because the boundaries of the relationship were not seen to be restricted by official time or limited to just the one project. This sense of solidarity was not developed by extensive communication or use of emails; instead, meaning was found in limited but relevant interactions, where expressions of commitment (like spending time) and care (as in sending gifts) transcended the complexities of the physical limitations of language. What magnifies the positive aspect of the Japanese ties is the contrast with prior experience of working with the North Americans, where relationships are seen to be rather superficial and set within business-defined boundaries. Quotations from Indian managers and developers expressed strong social ties with the Japanese, despite the long time it took to transcend initial boundaries:

I think in Japan those relationships matter that are considered sustaining. In the relationship it is primarily a very high degree of trust that the other person is not taking you for a ride. If he says something and I don't understand, I have faith in him enough and I go to him and ask him...to explain. Maybe he has a good reason for doing what he is doing. So that's the relationship when it is established. That is what sees us through all hot spots.

It is very easy to interact with the Americans rather than Japanese. You should be very careful, very polite with the Japanese. But once trust is built up there is nothing to [fall back on]. You just work on trust. In America, building up the relation has no value. In Japan, everything about the project is my own, like my own child. I have in so many instances made mistakes, but because of the rapport, it was okay.

The Japanese are very formal, very cooperative, and the person there tried to help and he asked us and came shopping. The first day at the airport, it was a big problem and they took us to their canteen. We could hardly get any food, and they stayed with us, and they went around showing us various foods. They took real pains to see that we had some food we liked.

The Japanese are very sceptical to begin with. They will scrutinize you, but once the relationship is built up, then they always rely on you. After a project, they will say: 'I prefer if you do the next project.'

And they just send the email and say that your group should take it up. This is because of the kind of trust and the rapport that is built up.

I found them very polite and very, very, sympathetic and helpful. Even in personal relationship development. Forget about the business. All my clients send me personal gifts. And it is not only to me, but also to my other group members. When I had my baby I received gifts from almost all my Japanese clients. Someone sent me CDs and songs, etc. So this is beyond your work, and it makes you feel it is more than work when you share a personal note. So I enjoy and relish most of my projects that I do there.

They are trying to be very friendly with us, and not being very official. So, I have a very soft spot for the Japanese. I can also give one more instance when I went to Japan last year to install something there. One weekend was free, I was so touched: the manager and the developer said: 'I have arranged [for] a man who knows English, to take you around Nagano, to show you the places around Nagano.' They took me for two days all around Japan. This is unlike what any other clients would have done. With North Americans, their relationship stops at 5 o'clock. Of course they help you, like when you have the problem of ticket bookings. Probably we Indians think too much about such personal things.

The Japanese are much more demanding, they have longer hours, much tighter schedules, and on the face of it they are stingy with praise. The USA is very different; when they come here they take the engineers out for a party and give them T-shirts and caps. They will say, 'You have done a fantastic job, etc.'. A Japanese equivalent would be 'Thank you for completing the job'.

The strong social ties with the Japanese are especially interesting because they have developed in conditions where everyday communication is problematic because of language, limited use of ICTs, different working styles and quite contrasting social structures. In all the Indian organizations we studied, we found some senior people who shared a passionate feeling towards Japan and were acting as champions to build up business there. These strong social ties seem to provide the strength to deal with the everyday communication challenges and to develop the relationship in the future.

9.4 Analysis of cross-cultural communication issues

The Japan–India case study provides an interesting study of contrasts in communication practices and styles, and their implications for the GSA relationship. We now present an analysis of these communication issues based on three key concepts:

- Lam's distinction of 'organizational' and 'professional' knowledge helps us to appreciate the *historical and social structures* that shape both the Indians' view of the Japanese being 'typically Japanese' and the contrasting Japanese view of the 'Westernized Indian'.
- We discuss how these different orientations to knowledge and the manner in which it is developed and shared have *inherent asymmetries* that give rise to challenges and responses to communication. We analyse these challenges using Hall's ideas that relate knowledge, information and meaning as a basis for developing a strategy for communication.

• The *structurational framework* already discussed sensitizes us to the fact that these asymmetries are not fixed and given, which can serve as mechanisms to explain change (or the lack of it). In and through the act of communication, these structures are being constantly challenged and negotiated, and new structures are being 'achieved'.

Guiding frames of reference and knowledge orientation

Lam's (1997) distinction of 'organizational' and 'professional' knowledge serves as a useful starting point to analyse the Japanese and Indian frames of reference. Although a lot of literature and experience about Japanese management classifies their working approach as being based on 'organization knowledge', there is relatively little knowledge about Indian companies. A popular interpretation of Indian organizations based on 'Hofstede-type' cultural dimensions emphasizes the significance of family-based networks, the role of hierarchy in decision making and the high level of aversion to risk taking and innovativeness. Structures that shape knowledge orientation in Japanese and Indian firms are extremely paradoxical; they reflect similarities and yet also marked differences. Both are Asian countries, with some congruence in values relating to respect for elders and family. There are, however, important differences. Khare (1999) contrasts Indian and Japanese work patterns; he argues that business customs in Japan emphasize the needs of the company over the needs of individuals, when relationships have to be preserved. A head office will support a subsidiary by giving it software development projects, even though the economic cost–benefit ratio was markedly less than in India or China. Khare writes that, in general, the sanctity of relationships, although being revered at home, is largely ignored in the Indian public setting. The concept of 'tatemae' which Hall (1984) describes as sensitivity towards others, or the public self, exists within the boundaries of the home and the immediate and extended family, rather than reaching the realm of the company or the nation. Khare's argument suggests that Indian firms are based on an 'organizational knowledge' where skills accumulation takes place in an internal labour market in which social relationships are significant. We argue, however, based on the case studies in this book, that the Indian firms engaging in GSAs have a radically different culture from traditional Indian organizations: the GSA firms reflect a much stronger global orientation and are often based on practices and methodologies from the West and North America. These firms are not based on an 'organizational' knowledge approach, as described by Lam, but reflect more of a 'professional' knowledge model, albeit a hybridized one.

Although we found that Indians see Japanese to be 'typically Japanese', the Japanese saw Indians to be 'Indian with a Western influence'. The Indians, especially the younger ones who are educated in computer science with a few years of experience in North America, have an initial perception of the Japanese as being poor in English, the spoken form much more than the written. They find the Japanese to be very detailed, taking very long to make decisions, and very ritualistic. Such judgements often translate to the negative view that 'the Japanese do not understand software'. As projects unfold, and the experience of working with the Japanese grows, the perceptions of some of the younger engineers are reinforced. Stories go around that Japanese take 14 hours to conduct a negotiation, or a team of 40 people came for negotiations, or that they are obsessed with quality and line-by-line code reviews. The Indian response, especially among the younger Indian engineers, is to revise their communication strategies in a functional way: by making sentences very short, or writing at the end of the message 'Please feel free to contact me if you need more information'. The more subtle and reflective staff analyse the various nuances, subtleties and differences across situations, companies, projects and technologies. They move beyond the stereotypes of 'Japanese do not like text' to 'Japanese managers in Firm X prefer graphics to text during requirements analysis'. The challenges of communication seem to be offset by the growth of strong social ties as the warmth, personal touch and depth of the relationship becomes appreciated. This appreciation is magnified when contrasted to the experiences of rather cold and business-like relationships with North Americans.

The Japanese perception of 'Indians with a Western influence' is guided by what they see as the Indians' professional knowledge model of software project management based on structured and formal methodologies that emphasize documentation and process milestones. Typically, the education of Indian engineers is grounded in American and European management styles, especially in computer science where the methodologies learned originate from North America. This orientation to Western education is further reinforced by their organizations' emphasis on CMM and ISO certification processes, that have their origins in North America. This 'Western orientation' is in contrast to the Japanese system where students primarily use books that help them understand their own management styles and work environment (Khare 1999). We found the Japanese perceptions of 'Indians being too Westernized' and 'insisting too much on formal documentation like the Americans, while we believe in discussion' quite strong. These perceptions potentially influence the decision of whether a Japanese firm will go to India or China for development work, since the Japanese view the Chinese as sharing similar communication styles. We saw some of the Indian firms becoming sensitized to such issues, as reflected in this quote by an Indian manager: 'We are trying to modify our software development methodology and give it a "Japanese flavour" for the future.' However, examples of such reflexivity are still rather limited in Indian firms and the sheer weight of the use of structured development methodologies and the North American experience seems to impede these reflexive attitudes and cause radical redefinition of project processes. These different knowledge orientations have some inherent asymmetries which have implications for communication. The asymmetries are primarily in terms of the different degree of tacitness of knowledge and the associated difficulties in communicating and interpreting it. We now analyse these issues, drawing upon Hall (1984).

Knowledge asymmetries and implications for communication

An interesting conceptual frame to examine communication issues arising through tacitness of knowledge is provided by Hall (1984). He argues that despite having made rapid advances in computers, effective linguistic translations are still not possible. The problem is not in the proper analysis of syntax and grammar, but in the relationship of the linguistic code to the larger setting of the scientific field that describes the context in which each word, sentence and paragraph is set. He argues that a communication transaction is affected by the cultural context in which it occurs, and this 'contexting' is an essential aspect of any communication and the meaning it conveys. Hall examines this process of 'contexting' through the functional relationship of information, meaning and context. Writing in *Beyond Culture* (1976) he gives an example of a couple living together for many years. When one of them comes home from a day of work, without a word being spoken, they can both understand very well what kind of day the other person has had. In contrast, in the court of law, nothing can be taken for granted and everything must be spelled out and a lot of information provided. There is thus a fundamental relationship between information, meaning and context that needs to be managed in developing a *strategy for communication.* To give people too much information is to 'talk down', and to give too little is to mystify them. In high-context situations, meaning is conveyed with minimum words while with low-context situations, detailed information has to be provided with specific words to make sense of the message. Information in high-context situations is provided not so much with words as through context.

Rules of communication vary from culture to culture. Hall gives an example of a North German who places a high value on doing things right and takes detailed, meticulous and low-context approach. While learning a foreign language, she/he will take pride in speaking correctly and following the rules of grammar exactly. It comes as a blow when a high-context Parisian corrects her/his French even though it is grammatically correct. Hall suggests that sometimes the rules of communication are embedded in advertising. A Rolls Royce advertisement may just say 'enough' while a Mercedes description would be packed with information, including a lot of performance statistics. Based on many years of empirical analysis and practical consulting, Hall has analysed why North American businessmen have a difficult time operating in Japan, as reflected in the common complaint that 'we don't know what the Japanese are getting at'.

In Hall's analysis, Japanese represent the high-context tradition where they do not get to the point quickly and provide a lot of contextual information. The North Americans, in contrast, take an approach that is based on logic, where a systematic analysis based on rules is the guiding basis. Inherent asymmetries are created because of this different logic, and these can give rise to different communication challenges. The Indian GSA firms represent a different breed than the traditional organizations that adopt more of a 'Japanese-like' organization knowledge approach. Operating in the global environment, and strongly oriented to the Western marketplace, the Indian firms operate with a toolkit of strong professional knowledge certified and standardized by global institutions. Such professional knowledge emphasizes documentation and a focus on a process that specifies milestones, deadlines and deliverables. The problem of attrition in the software industry means that the tenure of individuals in organizations is often limited; organizations try to manage this problem and to reduce the dependence on individuals by resorting to different kinds of computer-based knowledge management systems. They use techniques of 'shadowing' and 'backing-up' employees to reduce the shock of disruption resulting from the departure of an individual.

The communication challenges raised through the asymmetries described above are met in various ways, from the selective use of ICTs to the simplification of sentence structure, predetermined use of particular phrases in a message, the use of graphics versus text and the minimization of the transaction volume. Operating within a 'high-context' situation, the Indians, who professionally favour a 'low-context' orientation, are forced to revise their method of communication by reducing the amount of words they use, and by being very conscious of the meaning they convey through both structure and content. These are conscious revisions to a communication strategy that previously relied on more verbal and written communication, probably associated with lesser richness of meaning. The US manager sending long congratulatory emails or having elaborate parties with caps and T-shirts to mark the completion of a project is in stark contrast to the Japanese responding to a successful project completion with just a brief 'thank you'. Although the information conveyed is much less in the Japanese message, more meaning is associated with it: to the Indians it conveys a high degree of happiness, satisfaction and warmth. The Japanese 'thank you' could reflect an endorsement of future projects being given to the Indians without any questions being asked. In contrast, the North American manager, despite the effusive celebration, could quite possibly cancel the next project if he can get it more cheaply completed in the Philippines than in India. This may be less likely in a Japanese relationship.

Through this constant revision of knowledge of context, information and meaning, both sides are potentially revising the *structures of understanding* that shape their communication processes. As experience of working together develops and the initial boundaries of uncertainty and experimentation are crossed, social ties seem to grow extremely strong roots to provide the basis for radically redefining the communication processes. In and through the process of communication, we thus argue, the structures of knowledge are being constantly renegotiated, providing the potential to reshape communication processes. This structurational process is now described in more detail.

The structurational process

Structuration theory provides a conceptual frame to reciprocally link the micro and macro levels of communication. Through repeated action, patterns of interaction become established as standard communication practices in social systems that are reified over time. Drawing on a structurational perspective, Orlikowski and Yates (1994) have described communication as 'an essential element in the ongoing process through which social structures are produced, reproduced, and changed' (1994: 541). The structurational process, of which communication is at the heart, involves an ongoing process in which managers draw on the rules and resources for communication. Through the instantiation in communicative action, these rules are redefined. Here lies the core of our argument against functionalist and deterministic accounts of culture that treat cultural variables as causal agents for explaining change. We believe that the interpretations that people have, even if they promote national stereotypes, cannot be ignored as they are important determinants of action. We believe, however, that through the process of action, actors can revise their understanding of what they should or should not do. A static view of culture is fundamentally flawed as it ignores the implications of what people do in the everyday working of the GSA and how that influences their understanding. In Giddens' terminology, it ignores the fundamental *capability* and *reflexivity* of human beings.

This structurational analysis has interesting implications for the debates on globalization and the thesis of homogenization versus hybridization and localization. In chapter 2, we argued for hybridization as contrasted to either homogenization or localization. The question remains how hybridization takes place in situated circumstances, rather than whether or not globalization is about homogenization or hybridization. The structurational analysis of communication provides us with insights into the mechanisms by which processes of hybridization can take place. This is because GSA relationships cannot be conceptualized in an exclusive one-to-one state (such as India–Japan or India–North America) but as a *network of relationships* (India–Japan–North America). The same Indian firms and individuals are working or have worked with North American clients and are now doing so with Japanese firms. They bring meanings, understandings and methodologies from one context to the next and vice versa. An example of this is the manner in which software development methodologies are taken from one situation to another, adjusted, redefined and then re-exported. Communication helps us to understand these processes of hybridization in communication, and the manner in which the language and practices around it are taken into a context (such as formal and detailed specification documents), the processes by which they are resisted (Japanese complaining of 'too much documentation') and the manner in which methodologies are being redefined (with a Japanese flavour).

Another implication of our analysis is the use of the electronic medium, the Internet above all, which is popularly assumed to be able to normalize cross-cultural differences in communication. Such a perspective disregards the social and historical embeddedness of technology, even though the Internet is a communication medium with its own logic and language and not confined to one particular area of cultural expression (Castells 1996). There is always a historical and social specificity of the communication system and the use of it by people to do everyday things. Communication is embedded in social practice. Because the Internet is a medium for cultural expression and the encoding of ambiguity, it opens up a diversity of interpretation far greater than mathematical and formal reasoning. Castells' (1996) description of the relationship between new ICTs and social orders as 'problematic' has interesting implications for our case:

The inclusion of most cultural expressions within the integrated communication system based in electronic production, distribution, and exchange of signals has major consequences for social forms and processes. On the other hand, it weakens considerably the symbolic power of traditional senders external to the system, transmitting through historically encoded social habits, religion, morality, authority, traditional values, political ideology. Not that they disappear, but they are weakened unless they recode themselves in the new system, where the power becomes multiplied by the materialization of spiritually transmitted habits. (1996: 406)

The Japanese firms have shown a relative reluctance to embrace the new ICTs in the course of working with GSAs. This can be seen as one of the reasons why Japan in general has not wholeheartedly taken on board the idea of GSA, despite its potential and the realization that they need to respond effectively to demands for software development. The historical structures of employment, family and education, all contribute to shaping attitudes and perceptions about working with people from different countries, or using new ICTs. These structures are being challenged and placed under pressure to change in a variety of ways. In the GSA context, the next few years will be crucial for determining whether the Japanese enter more meaningfully into the 'GSA network'.

BIBLIOGRAPHY

Anderson, B. (1983). *Imagined Communities: Reflections on the Origin and Spread of Nationalism*, London: Verso

Avison, D. E. and Myers, M. D. (1995). Information systems and anthropology: an anthropological perspective on IT and organizational culture, *Information Technology and People*, 8, 3, 43–56

Castells, M. (1996). *The Rise of the Network Society*, Oxford: Blackwell

Chesebro, J. W. (1998). Distinguishing cultural systems: change as a variable explaining and predicting cross-cultural communication, in D. V. Tanno and A. Gonzalez (eds.), *Communication and Identity across Cultures*, London: Sage, 177–92

Couch, C. J. (1989). *Social Processes and Relationships*, Dix Hills, NY: General Hall

(1996). *Information Technologies and Social Orders*, New York: Aldine de Gruyter

Cusumano, M. (1991). *Japan's Software Factories: A Challenge to US Management*, New York: Oxford University Press

Edelstein, A. S. (1983). Communication and culture: the value of comparative studies, *Journal of Communication*, 33, 302–10

Ein Dor, P., Segev, E. and Orgad, M. (1993). The effect of national culture on IS: implications for international information systems, *Journal of Global Information Management*, 1, 1, 33–44

Hall, E. T. (1976). *Beyond Culture*, New York: Anchor

(1984). *The Dance of Life: The Other Dimension of Time*, Garden City, NY: Anchor Press/Doubleday

Hampden-Turner, C. and Trompenaars, F. (1993). *The Seven Cultures of Capitalism*, London: Judy Piatkus

Hofstede, G. (1980). *Culture's Consequences: International Differences in Work Related Values*, Beverley Hills, CA: Sage

Khare, A. (1999). Japanese and Indian work patterns: a study of contrasts, in H. S. R. Kao, D. Sinha and B. Wilpert (eds.), *Management and Cultural Values: The Indigenization of Organizations in Asia*, New Delhi: Sage, 121–38

Knoll, K. and Jarvenpaa, S. L. (1998). Working together in global virtual teams, in M. Igbaria and M. Tan (eds.), *The Virtual Workplace*, Hershey, PA: Idea Group Publishing, 2–23

Lam, A. (1997). Embedded firms, embedded knowledge: problems of collaboration and knowledge transfer in global cooperative ventures, *Organization Studies*, 18, 6, 973–96

Liebes. T. (1988). Cultural differences in retelling of television fiction, *Critical Studies in Mass Communication*, 5, 277–92

NASSCOM (2000). *The Software Industry in India – A Strategic Review*, New Delhi: NASSCOM

Nonaka, I. (1994). A dynamic theory of organizational knowledge creation, *Organization Science*, 5, 14–37

Ono, K. (1998). Problematizing 'nation' in intercultural communication research, in D. V. Tanno and A. Gonzalez (eds.), *Communication and Identity across Cultures*, London: Sage, 193–202

Orlikowski, W. J. and Yates, J. (1994). Genre repertoire: the structuring of communicative practices in organizations, *Administrative Science Quarterly*, 39, 4, 541–74

Schein, E. H. (1984). Coming to a new awareness of organization culture, *Sloan Management Review*, 25, 2, 3–16

Sharp, H., Robinson, H. and Woodman, M. (2000). Software engineering: community and culture, *IEEE Software*, January–February

Shewell, C. (2000). *Good Business Communicates Across Cultures: A Practical Guide to Communication*, Bristol: Mastek Publications

Shore, B. and Venkatachalam, A. R. (1995). The role of national culture in systems analysis and design, *Journal of Global Information Management*, 3, 3, 5–14

Triandis, H. C. (1972). *The Analysis of Subjective Culture*, New York: Wiley

Walsham, G. (2001). *Making a World of Difference: IT in a Global Context*, Chichester: Wiley

Westrup, C., Jaghoub, S. A., Sayed, H. E. and Liu, W. (2001). Taking culture seriously: ICTs, culture and development, paper presented at *IFIP 9.4 Conference in Bangalore* (2002)

Wynn, E. and Katz, J. E. (1997). Hyperbole over cyberspace: self-presentation and social boundaries in Internet home pages and discourse, *Information Society*, 13, 4, 297–328

10 Reflections and synthesis on theoretical insights

10.1 Introduction

The set of case studies discussed in this book emphasizes the significant complexity inherent in GSAs, and the need for theoretical approaches that allow us to go 'behind' the surface of the phenomenon and build insights that help answer questions about 'Why does the GSA process unfold a particular way?' In the cases discussed, the complexity of GSAs has been described as a multiplicity of inter-connected influences including aspects of industry, national economy, organizational strategies and the desires, expectations and mobility of individual developers operating in a global marketplace. The increased interest in GSAs has introduced new actors into ongoing debates on immigration and foreign policy. While firms may argue for GSAs based on economic and resource considerations, worker associations and unions criticize the potential danger of local job losses and argue instead for the retraining of unemployed adult populations. Government policy makers, especially in developing countries, face the complex task of balancing the need for fostering growth, which inevitably requires the active engagement of private sector investments that GSA firms provide, with the protection of rights of marginalized groups.

Beck (2000) uses the metaphor of 'Brazilianization' to describe the global employment market in the West which he considers to be increasingly characterized by unemployment and underemployment. Writing from primarily a European perspective, Beck paints a rather bleak picture of the world of work, and tends to downplay how global relocation of work from countries such as Germany can lead to opportunities in other parts of the world, such as India. New technological developments and the opening up of new global markets as a result of liberalization processes tend to destabilize ongoing relationships, as the potential to establish new GSAs emerges. GSAs can thus be seen as an interesting instance of the global employment market that Beck describes, as they bring together a complex set of issues including politics, economics and social and cultural change. German policy on the 'green-card' scheme to attract non-European software programmers, for instance, has helped to open up broader debates on policies towards foreigners and immigration. These debates are intricately linked to questions of how and with whom German organizations should form global alliances, which in turn has implications for individual-level issues of who gets what kind of jobs and the skills that become important or redundant as a result. These questions in turn have further implications for other systems such as education. Should English be used to teach courses that have historically been taught in the German language, for example?

This complexity inherent in GSAs cautions against taking single-level and discrete approaches that seek to identify factors that influence the decisions surrounding outsourcing. While sensitivity to these factors is important in appreciating important issues, they are limited in explaining the inter-connectivity between such issues and how they are redefined over the duration of the GSA relationship. We have conceptualized GSAs in a 'model of' and 'model for' relationship with processes of globalization, which implies that GSAs reflect, and also serve as carriers of, globalization. While GSAs reflect the 'network society' structure, understanding the working of GSAs helps us to develop insights into the dynamics of networked and global organizations. The 'successes' or 'failures' of GSAs lead actors to develop positive or negative attitudes towards not only globalization but also the countries and organizations involved. As organizations enter new countries, managers need to learn the local 'language', and develop a broad cultural understanding of their alliance partners' attitudes towards such issues as time deadlines and quality. Even though these forays into culture are initially often very superficial, they potentially provide a point of entry from which actors can develop deeper cross-cultural understandings.

The case studies in this book have traced the mutual linkages between GSA processes and globalization through the different theoretical lenses of standardization (chapter 4), identity (chapter 5), space and place (chapter 6), knowledge transfer (chapter 7), power and control (chapter 8) and cross-cultural communication (chapter 9). A primary focus in each case has been on one key issue (for example, standardization). This individual focus raises the danger of an analysis that can be criticized for being 'reductionist', but it permits an in-depth theoretical study of a particular issue. Our intention is not to argue that one issue 'corresponds' to a particular case, but to find an appropriate balance between exploring an issue in depth and a synthesis across the issues. While not denying the inter-connectivity that exists across the issues, we draw attention to the themes that became prominent during data collection and its analysis, the reading of related literature and theory-building. For example, we have used standardization as the primary lens for the analysis of the Witech case in chapter 4, while acknowledging the role standardization also plays in shaping complexities in knowledge transfer and in the construction of identity. The separation of these themes has been made primarily for analytical purposes. The aim of this chapter is to shift the focus to the inter-connectivity between the themes and develop a synthesis of the various theoretical issues. Finding this balance will help us respond to Walsham's (1998) call for the use of 'theories that can help us to generalize the results from micro-level studies' (1998: 1081).

Before we discuss the synthesis of the theoretical themes, we reflect on how the theoretical ideas evolved. The aim of this reflection is to provide some insights into the process of theory-building, especially the relation between empirical data and conceptual ideas.

These reflections are discussed in the next section, followed by a more detailed discussion linking up the themes in the book. Finally, we reflect on some limitations and possible future extensions of the research programme.

10.2 Reflections on the theoretical process

A key source of inspiration for our theory building was Castells' well-cited trilogy (1996) that articulated a theory about new social structures in the late twentieth century. Castells' concept of the 'network society' helped to introduce an informed debate on the meaning of the multi-dimensional transformations that are taking place in contemporary society. More recently, Castells (2001) argues that the *concept* of the 'network society' should replace the *notion* or *metaphor* of the 'information society' in aiding our understanding of contemporary social structure. The important distinction Castells makes between metaphors, concepts and theory is elaborated in this extended quote:

> In some cases, the use of metaphors allows one to indicate a line of interpretation without closing the meaning, in an old epistemological tradition suggested by Bachelard. Yes, at the end of the process of theory building, we should have a clearly defined system of unequivocally defined categories. I concede without any problem that some (but not most) of the categories I use are metaphorical, and evocative, rather than conceptual because it is too early to close their meanings... Unless and until a systematic theoretical construction is able to produce knowledge when confronted with observation, for me it is not theory, it is discourse, ultimately either a formal game or literary text. So, I prefer the uncertain, less elegant path of approximating categories and observation, and adjusting metaphors to data until they become concepts as tools of grounded theory. I believe this is a fundamental, epistemological option that may take social science away from the impasse of metaphysics without surrendering to empiricism. (2001: 17)

Castells argues for a 'theory in practice' where the relevance of a theory is judged by its ability to yield understanding when the phenomenon is subjected to observation. While metaphors and evocative notions like the 'information society' help to provide an initial language to describe social structure, they have limited explanatory potential when subjected to empirical observation. Castells argues that replacing a description conveyed in a notion or a metaphor by a 'concept' is not just a terminological matter, since a concept proposes a certain meaning and pretends to be based on observation. Castells' 'network society' is such a concept about the nature of contemporary social structure and is developed and examined over a period of sustained and systematic empirical observation and analysis. Castells' 'theory of practice' perspective resonates with van Maanen's (1989) description of the process of theory-building using the metaphor of an ongoing 'conversation' between empirical data and our theoretical ideas and frames of understanding.

The ideas of Castells and van Maanen guide our approach to building a theoretically informed empirical analysis of the process of GSA relationships. We started this research in 1996 inspired by Giddens' (1990) writings on globalization and modernity. Specifically, we were interested in examining how Giddens' notions of institutional reflexivity, disembedding mechanisms and time–space distanciation could be useful to examine the relationship between GSW and the ICTs used to support and coordinate this work. Our research started as we obtained access to the ongoing GSA of GlobTel with four Indian firms, a relationship considered pioneering in the Indian software industry and globally. Since this relationship was established in the late 1980s, it has provided access to a rich history and learning about four individual GSAs, thus enabling a higher-level analysis through inter-case comparisons. The extremely dynamic period the industry was enduring also permitted us to observe closely the trials and tribulations that firms suffer and their responses in such turbulent conditions.

In the following paragraphs, we discuss the processes that evolved in developing our different theoretical ideas.

Space and place

Over time, as our understanding of the empirical issues developed, we started to explore various theoretical ideas through 'conversations' with other academic disciplines and by the reading of relevant literature. Given the fundamental role of space and distance in GSAs, studies in human geography provided a rich source of ideas to analyse the relationships between material practices and the social and physical domains in which they occur. We were struck by the contradictions and tensions between the GSA assumptions that emphasize distance and 'space', and individual needs for proximity and 'place'. These tensions varied with the different stages of the relationship. The ongoing, and sometime self-destructive, nature of these tensions made us explore the concept of 'dialectics' and the manner in which it has been discussed in philosophy, human geography, organization studies and, in relatively less depth, in IS. The combination of the concepts of dialectics and space/place provided the basis to analyse the GlobTel–MCI case. The theoretical framework of the 'dialectics of space and place' that developed through this rigorous interaction between empirical analysis and conceptual ideas potentially contributes to various ongoing debates in IS and globalization. For example, while Schultze and Boland (2000) describe the activities of IT contractors in an American firm as a dialectical arrangement over space and place, they do not elaborate on issues related to how to manage this process. Our theoretical framework can potentially add further insights into the dynamics of this process. For example, the actors' need for identification with particular 'places' is in a dialectical tension with their job demands of operating in a variety of 'spaces', with significant implications for identity.

Identity

Giddens emphasizes the inter-relationship between institutional-level dynamics and self-identity as an important feature of current globalization processes. Similarly,

Castells describes the 'power of identity' as a defining aspect of the network society. In contrast, our interest in this concept was at the organizational level and how it was drawn upon by managers to shape GSA processes. We encountered managers, for example, those in ComSoft, who explicitly discussed identity, and its practical implications for everyday processes such as the management of attrition. However, identity was not static and was continuously enacted and redefined in action. The mission of 'unleashing Indian creativity' was initially a defining aspect of ComSoft identity, held together by the ideological glue of the ethos surrounding 'made in India'. As discussed in chapter 5, further fieldwork in Japan emphasized the managers' preference for technology approved in the USA, which led to ComSoft needing significantly to reposition both identity and image. The speed at which these dynamics unfolded made us reflect on the largely functional role of identity, and the conscious manner in which it was continuously hybridized in response to changing global conditions. The idea of boundaries, external and internal, and the manner in which they are continuously negotiated in GSAs, was important in appreciating the image and identity linkage and the role of organizational culture in shaping this relationship. We draw upon Giddens' (1984) structuration theory to develop an interpretative framework to conceptualize organizational culture in terms of rules and resources that actors draw upon in shaping agency in the construction of identity and image. This structurational analysis helps to move some of the debates in globalization debates to a more practical level. This also helps in addressing the question of how the 'power of identity' unfolds in particular micro-level situations of GSW in particular, and global work more broadly. The 'power of identity' is observed above all in the response of actors to attempts to standardize their everyday work practices.

Standardization

Towards the end of the period of the empirical work on the GlobTel relationship, one of the authors moved from Canada to Oslo where a strong research tradition existed focused upon technology, standardization and actor-network theory. Ongoing conversations with colleagues, further readings, attendance at seminars, listening to PhD student presentations and various informal discussions inspired us to examine the data with the 'lens of standardization'. These discussions and readings enhanced our awareness and sensitivity to the role of standards, standardization processes and the tensions that ensue in establishing and maintaining a large infrastructure to support a GSA. Such concepts provided a powerful way to develop a *holistic perspective* on the relationship and emphasized the multiple and complex socio-technical and political processes that evolve over time to establish standards and gain compliance. Our analysis helps to extend these existing debates on standards and standardization to the domain of management practices, thus broadening the earlier primarily technical focus. A focus on how standards unfold through processes of use also helps to extend the previous theoretical discussions which had focused mainly on the politics of how new standards were created.

An insight into these additional dimensions to the standardization question potentially yielded deeper understanding into the 'global–local' debate. Management practices can be seen as more closely related to 'local'-level issues that are relatively more complex to standardize than technology-related questions that can be conceptualized as more 'global' issues, relatively more amenable to standardization. Issues of standardization are linked to issues of power and control, especially relating to who has the power to standardize and how.

Power and control

The theoretical ideas in this area developed during the course of the doctoral studies of one of the researchers and his travels to India on holiday to meet the other authors of this book. The interest that the doctoral researcher developed in India provided the motivation for the empirical work on his PhD that was carried out in the UK and India. Initially, the researcher experienced a sense of frustration and even anger after the initial interviews as he felt that the Indians were being treated in a rather instrumental manner by the UK firm, echoing the historical colonial pattern. This led him to undertake further reading relating to Indian culture and society and provided the basis for the initial political and cultural analytical framework set within the context of globalization. Early reading of the data was sensitized by Hofstede's (1980) analysis of cultural consequences, which was soon found limited for the various reasons discussed in the case studies in earlier chapters. Subsequently, Morgan's (1986) metaphors of organizations, specifically the political and cultural metaphors, provided a useful lens for the analysis, which primarily focused on new forms of power and control emerging from the impact of globalization.

The first attempt to publish the work in a journal led to many useful suggestions from reviewers who encouraged the use of structuration theory for the analysis. Another paper (Sahay and Walsham 1997) that had analysed a GIS implementation case in India using a structurational lens provided a further basis for the development of a revised framework. Particular attention was given to theorizing the issue of 'control at a distance', a crucial concern in GSAs. This process of theorizing was aided significantly through 'critical discussion groups', in which one of the authors was a member, along with PhD students and colleagues at Manchester University. These various discussions and readings became the basis for the resubmission of the paper and its subsequent publication in *Information and Organization* (Nicholson and Sahay 2001). In revising this paper in line with the aims of this book, we consciously tried to situate the analysis in relation to phases of the GSA viewed within the globalization context. Foucault's work was inspiring in the analysis of the power–control–culture relationship, and this was incorporated into the theoretical framework. Comments by an organization specialist led to a more focused discussion of the case study data in relation to the revised theoretical frame. Foucault's ideas that emphasized the power–knowledge relationship also have implications in appreciating the complexity of knowledge transfers.

Knowledge transfers

The knowledge transfer analysis has its root in the Sierra case presented in chapter 7. What was interesting about the case was the manner in which Sierra established their India operations full of hopes and expectations, and how they shut down some two years later full of remorse and disappointment. We were interested in understanding why this happened, and a closer examination of the data emphasized the difficulties experienced in knowledge transfer, and the practical realities of how the process differed very significantly from their expectations. An initial analysis of the case was presented in the *IFIP 9.4* conference in Cape Town (Nicholson, Sahay and Krishna 2000) drawing broadly upon Giddens' (1990) analysis of modernity. Suggestions from the audience at the conference broadly pointed to the importance of knowledge-related issues in the case. Initially, we were guided by Blackler's (1995) 'images of knowledge' because they went beyond the objective perspective taken to 'knowledge management'. We also explored related literature on interpretivist approaches to knowledge drawing from philosophy, management and sociology in addition to IS. While we found Blackler's images to be useful in sensitizing us to the 'different kinds of knowledge' involved in a GSW, we felt the images metaphor was limited in developing the process perspective that this book was trying to articulate. We found the 'community of practice' (Lave and Wenger 1993) perspective to be useful in trying to link Blackler's images of knowledge with the practice of GSW so as to develop the processual analysis. In development and refinement of this framework we were greatly supported by feedback and discussions on early drafts by PhD students and other colleagues at the University of Manchester working in the area of knowledge and knowledge management. The complexity in knowledge transfer is fundamentally linked to issues in communication, and how knowledge about products, processes and practices can be transmitted, interpreted and used by the different parties involved.

Communication issues

Towards the end of the empirical work in the GlobTel project, we found an increasing frustration in both GlobTel and the Indian firms about the future of the relationship, which seemed to be achieving a 'plateau state'. We were intrigued by an increased mention in the Indian firms, especially ComSoft, about the rich potential of Japanese markets especially when compared with North American clients like GlobTel. By this time, we had developed a high level of trust with the ComSoft management and felt comfortable in requesting permission to visit Japan and meet their office staff and also some of their existing and potential clients. Meetings with ComSoft developers in India working on Japanese projects, followed by interviews in Japan and Korea, helped to clarify some of the initial issues that both sides were experiencing in establishing GSAs. The Korean interviews served to sharpen the analysis as the kind of communication problems experienced there were similar to those in Japan.

We were struck by the challenge of communication that seemed to lie at the core of the definition of the relationship, from the kind of business model adapted, to the nature of projects outsourced and to shaping inter-personal relationships. Since we did not follow any one relationship longitudinally, but talked to people in different firms about their perceptions and expectations, we could not develop theoretical insights into the nature of the process. Instead, we obtained a 'snap-shot' of the factors that managers described as important in shaping their perceptions about the GSAs. We then related these issues to other writings on cross-cultural communication. Our contribution is in emphasizing the importance of *national stereotypes*, and their role in shaping communicative action. Our analysis presents a point of departure from typical academic debates that criticize the use of national stereotypes in studying culture on the grounds that it is reductionist and limited. We rather argue that these stereotypes are crucial to understanding, as actors draw upon them in shaping their everyday action. The stereotypes that the Indian developers have about Japanese, for example, are reflected in the structure and content of the messages they send. This *structurational perspective* emphasizes that these stereotypes do not remain static but are redefined and also strengthened in and through the process of communication.

Another interesting aspect of our analysis was in the comparisons that the Indians made concerning the North American and Japanese styles of communication and working. These comparisons were significant as they both consciously and unconsciously helped to provide a frame of reference for the Indians to make sense of their Japanese experience. This emphasizes the importance of situating the analysis of GSAs in a *network structure* and not just in one-to-one relationships that tend to obscure the larger picture. Since actors and firms are situated in a multiplicity of relationships in the network, experiences developed in one part inevitably have 'spread effects' to others.

While this discussion helps to describe the individual processes involved in the development of theory, we develop in the next section a synthesis that looks across these issues. This has three key components:

- It summarizes some *key features* of the themes, drawing upon examples not only from the particular case in which it had been discussed earlier, but also by looking across all the various cases.
- We then develop an *inter-theme* and *inter-case comparison* that permits us to see how some of these different themes relate to each other.
- We then discuss how the analysis of particular micro-level themes contributes to our understanding of the *macro-level social theory* such as that of Giddens, Beck and Castells upon which we have drawn.

We develop these contributions drawing upon the framework of the 'model of' and 'model for' relationship between processes of GSA growth and globalization. To aid the understanding of this rather detailed discussion we make use of multiple tables. Two sets of boxes are presented for each theme that first summarize key features and secondly relate these features to the 'model of' and 'model for' relationship.

10.3 Theoretical synthesis

We have identified and developed six concepts (also referred to as 'themes') within the framework of the 'model of' and 'model for' metaphor to examine the empirical relation between the processes of GSA evolution and globalization:

- Standardization
- Identity
- Space and place
- Complexity of knowledge transfer
- Power and control
- Communication.

We discuss each of these themes in the framework of the three aspects of the synthesis outlined above.

Standardization

Key features

We first summarize in box 10.1 the key features of standardization.

> **Box 10.1 Key features of standardization**
>
> - Takes place within and across the GSA at multiple levels of the organization, groups and individuals
> - Wide scope across physical, technical, managerial domains
> - Ongoing process of negotiation, resistance and redefinition surrounding the implementation of standards
> - Ongoing destabilization of standards by external events in the industry and global marketplace
> - Rising work levels require more sophisticated forms of standardization, and an associated hierarchy of standards
> - Systems of quantification play key role in the implementation of standards
> - Various explicit and implicit translation mechanisms used by managers to enable the implementation of standards

Standards are pervasive at multiple and inter-connected levels of the GSA relationship from inter-organization, organization, project, down to the individual. In the Sierra case, the management attempted to standardize at the inter-organization level, by creating a 'little bit of Sierra in India' that would identically mirror the UK operations at both the formal and informal level. The formal level included management practices such as using similar things as personnel appraisal forms in India and in the UK. At the informal level, attempts were made to construct in India a similar culture as in the UK: drinking and swearing were seen as active indicators of creativity, for example. The Gowing case (chapter 8) demonstrates how standardization can be played out at the organization level. Gowing UK consciously developed the 'sociological experiment' of bringing Indian programmers to the UK with the aim of introducing standardized methodologies for software development (and with it a discipline in their use) across the Gowing group of companies. Since many of the Gowing companies had been individually acquired, they reflected very different sub-cultures including styles for software development. A challenge for the Gowing management was to eliminate these differences and standardize the culture, in this case through the 'rock of discipline' that the Indian programmers provided. GlobTel's attempt to create the 'Global manager' (chapter 4) reflects a powerful attempt to standardize at the individual level by creating universal frameworks within which managerial action could be carried out and evaluated. Attempts to standardize accents of developers as described in the Sierra case reflect another complex individual-level attempt at standardization.

At these different inter-organization levels, the organization and the individual, the scope of standardization efforts is extremely wide and covers physical aspects, technical artefacts and managerial processes. Standards are never a given and an absolute, but are continuously negotiated, reinterpreted and redefined over time. These processes of implementation and negotiation can be seen as a series of translations, both explicitly and implicitly shaped by management action. Each translation involves the loss and gain of some part of the standard.

Inter-case and inter-theme comparisons

Standardization is related to a number of other themes, including knowledge transfer. During the early stages of the relationship it is *product-related knowledge* that needs to be primarily transferred. However, as work is upgraded, transfer of knowledge about *processes and practices* becomes increasingly important. Processes of globalization and GSW operate under implicit and explicit assumptions that knowledge can be seamlessly transferred across time, space and cultures. However as the cases in this book have emphasized, particularly the Sierra analysis, these processes of knowledge transfer are fraught with various complexities arising from the tacit nature of understanding, communication problems, varying competence levels and different socially constructed perspectives of knowledge.

Issues of standardization and knowledge transfer are also fundamentally linked to questions of *identity*. As work enters into more unstructured domains, largely dependent on the expertise and talent of key individuals, issues of identity become crucial. On the one hand, the evolution of work becomes dependent on whether these key individuals remain in the organization. On the other hand, as these individuals gain in expertise and knowledge, they become more attractive in the global marketplace. In trying to retain these individuals, companies such as GlobTel try to construct 'Berlin Walls' by inserting standard clauses into the contract to prevent movements of developers. The ways in which individuals respond to these restrictions are fundamentally linked to issues of identity and the issues that give them meaning and relevance – issues that are inherently 'non-standardizable'. For example, it is extremely problematic to standardize accents by 'neutralizing' them. Standardization is intimately connected to issues of *power and control*, as someone's standards are always being imposed over someone else's. This was highlighted in the GlobTel case where despite Witech's high-level standards of quality it was made to follow GlobTel's standards.

Contribution to social theory

Box 10.2 summarizes key inferences about the nature of the 'model of' and 'model for' relationship.

Box 10.2 'Model of' and 'model for' relationship: standardization and globalization

- Global separation drives the need for standards that create new conditions and perceptions of global separation
- Global separation implies the need for structured and standardized work that over time with learning in a global environment enters into relatively unstructured domains; this creates new need for standards
- Increasing diversity of global networks of GSAs by the entry of new countries and firms raises demands for new standards
- The interesting question while operating in these global networks, is 'What remains invariant in the process of translation?'
- We analyse this using the metaphor of 'scaffolding'

A key contribution arises from the analysis of the relation *between work and standards* reflected in the hierarchy of standards. During the initial stages of the relationship, when there is uncertainty about aspects of separation, the work externalized is often independent and relatively structured. To enable such work, standardization is focused primarily at the level of the physical and technical infrastructure such as buildings and the technical architecture. However, as work evolves and enters into more unstructured domains, processes become relatively interdependent. These processes increasingly involve issues of ownership and responsibility, both in letting go (by GlobTel, for example) and acceptance (by Witech, for example (chapter 4)). Under these circumstances, the standardization of management processes becomes crucial. Transfer of ownership raises the need for project management to be closely coordinated – by standardizing the holding of weekly videoconference meetings to monitor progress, for example. These new forms of standards not only reflect the changing work processes, but also require actors to deal with separation differently. Analysis of the dynamics of work and standards relationship helps to develop insights into Giddens' (1990) discussion on the role of *disembedding mechanisms* in globalization. While Giddens describes these mechanisms as being an important aspect of globalization, he does not elaborate on how these processes occur in situated practice. The micro-analysis of the structure of work and distance is an aid to understanding the role of standards in mediating this relationship over time.

Organizations, by virtue of the very global networks in which they are situated, reflect the *informational capitalism* that Castells (1996) describes as characterizing modern life. By virtue of these dynamic networks, GSA relationships are never stable: they require an ongoing process of revision and implementation of new standards. For example, as discussed earlier, Japanese managers did not appreciate the ISO 9000 type of methodologies used by the Indians. The global marketplace and the turbulent technological environment within which GSAs are situated make existing standards extremely vulnerable to destabilization and continuously raise the need for new standards. The proliferation of the Internet in the domain of software development for telecommunications, for example, brings with it the need for new tools and development methodologies situated within different frameworks of standards. Within these global networks, standards are continuously being changed or reinforced. The interesting question concerns 'What remains invariant?' The 'scaffolding' metaphor helps to examine this question and build insights into the nature of the evolution of the GSA process. The hierarchy of standards sensitizes us to the question of the risks that are associated with applying different standards. This aids our understanding of Beck's (1992) notion of 'redistribution of risk', since application of standards can be seen as 'risky business' despite the intention to eliminate both risks and uncertainty.

Identity

Key features

We first summarize in box 10.3 some of the key features of identity.

Box 10.3 Key features of identity

- Organizational culture, identity and image are inextricably intertwined and conceptualized through a structurational and self-referential process
- Agency that shapes identity can be seen as constructed through situated practice and mediated by its linkage with organizational culture and image
- Identity is shaped as a function of the membership actors have in multiple social systems, including Indian society, global high-tech business, the academic community and the firm
- Transformations in identity take place in an extremely dynamic manner that is magnified in a global and turbulent marketplace and industry; identity is crucial in shaping different stages of the GSA process
- Global networks in which GSAs are situated lead to a hybridization of identity

---

Features of identity concern the process of its construction, including the different sources of identification, and its role in shaping GSA relationships. Identity is constructed in structurational processes, shaped by the membership of managers in multiple social systems, including those related to society, high-tech business, the academic community and the firm itself. The structurational process contributes to developing agency that is directed both internally, helping to shape the biographies and careers of *individuals*, and externally, helping to define the *firm's image*. In the GlobTel case, Ghosh and Paul, despite having lived away from India for many years, still retained a strong sense of identification with Indian society. This identification became an important reason for the GlobTel programme being initiated in India in the first place. In the Sierra case, the country manager Mitra, despite having the same roots by birth as Ghosh and Paul, did not share a similar sense of identification and attachment to Indian society. There was thus a lower threshold level of tolerance and understanding in dealing with the complexities associated with working there. The nature of technology and associated work that finds its definition in the structure of the 'global high-tech' business provide another important source of identification. Developers, especially the younger ones, seek avenues where they can work with cutting-edge technologies and new products. Working with legacy-type technologies has negative implications for developers' self-worth and their sense of relevance in the global marketplace.

Identity plays a key role in shaping various stages of the GSA process from the inception of relationship to its growth and its stabilization or breakdown. A key thrust of the Sierra management was to create an identity of its India office that would mirror the UK operations. When it became apparent that 'sameness' was not possible, the resulting disillusionment was instrumental in closing down the operations in less than three years. In Gowing, a key initial motivation of working with India was to create a coherent identity in the UK operations. The Gowing CEO believed that the Indian developers could provide the 'rock of discipline' and coherence in a previously fragmented identity made up of the diverse sub-cultures that came as a legacy of acquisition. However, while the Indians provided coherence in identification with the UK office, they themselves often experienced an alienation and 'placelessness' when residing in the UK owing to feelings of loneliness and uncertainty about where they belonged. Their loss or fragmentation of identity often contributed to their departure from the organization. This caused problems in project management, as there was no buffer available on-site to deal with the contingencies of the loss of key staff. In the MCI case (chapter 6), as the relationship stabilized and MCI was left primarily with the burden of legacy work, the management felt that it had compromised its 'MCI' identity in trying to come 'too close' to GlobTel. This realization of the need to reassert its 'true identity' would be key in defining how the relationship shapes up in the future.

Inter-case and inter-theme analysis

Issues of identity are linked to processes of standardization. To prevent attrition, management standardizes HR practices, such as specifying the minimum number of years a developer should work, thereby curtailing their sense of freedom. The issue of knowledge transfer is also intimately linked to issues of identity. As in the MCI case, the developers in GlobTel and MCI had very different frames of reference, of telecommunications and software, through which their identity was defined. Telecommunications knowledge is gained and stabilized after many years, in contrast to the overall software industry trends where the focus is on acquiring skills (for example, Java or C++) required for a fast-changing marketplace. These different temporal frames within the industry were important in shaping identity and had implications for building differing interpretations of attrition.

Issues of identity are also connected with questions of power and control, particularly over whose frame of reference will be used as a 'template' in defining identity. GlobTel's 'global manager' initiative emphasized the power they wielded over the Indian firms as it defined its identity. However, these attempts are never without resistance and negotiation. As the Sierra case emphasizes, the attempt to develop a 'little bit of the UK in India' was never realized, which led to a disappointing closure of the India operations. The identity of developers is fundamentally bound up in issues of 'place', and 'space-like' issues have significant implications for identity. Over time, intimacy can be developed with different domains, such as the Indians feeling comfortable in Japan, and new and hybridized forms of identity can thereby be constructed.

Contributions to social theory

In box 10.4, we first outline some of the 'model of' and 'model for' features that help us to infer contributions to social theory.

Box 10.4 'Model of' and 'model for' relationship: identity and globalization

- Global imperatives of cost saving achieved through retention of bright developers, requiring individualized care; as a result, the constantly increasing costs of doing work are in conflict with cost saving imperatives through distance
- Managing the 'openness' and 'closedness' of the organization operating in the global marketplace is shaped by, and also shapes, questions of identity
- To prevent attrition in a global marketplace, management attempts to standardize HR practices, which in turn frustrates the worker and encourages movement
- Creating 'sameness' in knowledge workers conflicts with their quest for individual creativity
- Choices of workers are shaped within standardized 'templates' of reward and recognition defined within a capitalist framework of economic and political rules

---

For the proponents of the neo-liberal ideology that characterizes globalization, GSAs serve as a 'model of' this marketplace with developers and firms supposedly possessing 'free' choices on where and with whom they want to work. This global mobility of developers is in tension with the organizational need to retain the most talented programmers (classified as 'key resources') by providing them with individual attention and care. This individualized attention adds to the salary bill, and steadily conflicts with the cost saving objectives of distance work. While management draws on the aspect of identity in an attempt to retain and motivate staff, individuals try to affirm their identity and seek new opportunities. As firms operate in new global networks, they are implicated in ongoing revisions of identity. The linkage between culture, image and identity helps us to understand complexities inherent in this process of revision. The ongoing transformations of identities that occur in GSAs supports Castells' (2001) argument that social structures are not inevitable: they change depending on what social actors do in particular situations.

Analysis of the process of ongoing revisions of identity provides insights into the question of what Castells describes as 'resistance identity' and the conditions under which it becomes 'project identity'. 'Resistance identity' refers to the different pockets of resistance that arise in relation to globalization – for example, the protest against the WTO meeting in Seattle in 1998. Such resistance can take the form of a 'project identity' if it coalesces with other organizations and their forms of protest to create institutional structures that reflect new forms of governance. This process of conversion, Castells argues, is extremely complex and poorly understood. While Castells' argument is made primarily with relation to social movements, it has relevant implications for GSAs. For example, to deal with attrition (which can be seen as an expression of 'resistance identity'), firms may set up new structures for compensation and reward/recognition. Institutionalization of these structures provides the firm with a different basis for governance, and a new form of 'project identity'. The conditions and form in which this conversion takes place is an empirical question for further IS research.

Creating 'sameness' in knowledge workers as a strategy of management to stem attrition conflicts with their quest for individual creativity. This conflict relates to Beck's (1992) discussion on the tension between *individualization* and *standardization*. While individuals have a range of individual choices on where to work, these choices are shaped within standardized 'templates' of reward and recognition of a capitalist-defined economic and political framework. Breakdowns in these frameworks were reflected in the US dotcom crisis and further magnified by the 11 September attacks, thereby making individuals and firms increasingly vulnerable. Attempts to reassert identities have come with Indian firms trying to enter different geographical regions (such as Scandinavia) and new technological domains (such as e-commerce). This switching of identities and image is difficult to accomplish in practice because of historical and geographical realities. When Indian firms try to assert their presence in Scandinavia, a geographical area largely unknown to them, for example, they draw upon an identity shaped largely through their earlier North American experience. The Scandinavian managers often feel uncomfortable dealing with this expression of a North American style; they show instead a preference for Russian outsourcing firms with a more compatible 'European identity'.

In the discussion on identity, we extend the work of Castells and Giddens by reinforcing their emphasis on identity, and broadening the discussion to the business organization, a level previously largely ignored. Castells (2001) argues that work itself may be declining as a source of identity, even though people work longer hours and more intensely than before. Castells argues that consumption is not the dominant source of identity, because meaning is not obtained from the act of consumption itself but from some external cultural codes that give consumption meaning (gay culture or body piercing, for example). In GSAs, the primary source of meaning can be seen to come not from the work itself but more through the *cultural codes* of the global and high-tech business in which GSAs are situated.

Space and place

Key features

We outline in box 10.5, the key features of space and place.

Box 10.5 Key features of space and place

- Space and place serve as 'root metaphors' to analyse the relationship between social practices and the physical and electronic domains in which they occur
- These metaphors need to be integrated with the material and geographical realities of GSAs
- While space is an arena in which universalized activities occur, place is associated with particularity; space is associated with 'becoming' and growth, and place with 'being' and local experiences
- The dialectical interplay between space and place helps to shape social interactions and the nature of work, and with it the process of growth of a GSA
- Dialectics are not viewed as a deterministic logic but as an interpretive approach representing various arguments and concepts
- Principles of totality, change and contradiction help to interrogate the dialectical processes.

Given the fundamental role that geography plays in GSAs, space and place can be seen as 'root metaphors' to understand the relation between the work processes of GSAs and the physical and electronic domains in which they occur. Place is associated with proximity, particularity and 'being', but space represents distance, universality and 'becoming'. The metaphors of space and place, as Harvey (1996) argues, need to be integrated with the material and geographical realities of capital and distance. This integration is developed in the MCI case (chapter 6) through the analysis of the intricate and ongoing relationship between work, telecommunication infrastructure and space/place. While certain kinds of work can be done effectively in 'spaces', with the passage of time, and as stability is attained at a certain level of work, expectations change on the kind of work to be done. The changes in expectations create varying needs for proximity and inclusion, which are always *contested and negotiated*. We have argued that such tensions can be conceptualized in a dialectical frame of reasoning, where dialectics are not viewed in terms of a deterministic logic but in an interpretive sense consisting of a process of argumentation and concepts.

Inter-case and inter-theme comparison

The metaphors of space and place help us to gain deeper insights into issues such as standardization. To deal with conditions of separation, organizations develop various forms of standards. Firms often standardize project management practices by scheduling weekly videoconferencing meetings, for instance, on Monday mornings, the agenda for which is exchanged in advance on the previous Friday afternoon. Such standardization in the use of time, space, technologies and procedures, although helping to deal with project management complexities arising through separation, also helps to create new meanings of 'distance' among the actors. Excessive use of videoconference meetings can make developers feel as if they are being 'micro-managed' by their GSA partners, for example. This leads to resistance, and proposals for alternative control structures. Some organizations attempt to deal with distance by outsourcing only mature and structured technologies, the idea being that well-specified problems can be worked on in 'spaces' enabled by the use of standard software development methodologies. New technologies, however, require more 'place'-like understanding. With time, learning occurs about how to deal more effectively with distance in GSW. This learning allows a confidence and trust to develop in entering into more unstructured domains such as new product design. Working in new unstructured domains raises the need for new types of standards, for example relating to communication processes (making accents 'neutral', for example) to help avoid the problems of (mis)interpreting requirements, especially over the electronic media of phone and videoconference.

The assumption that place does not matter in GSAs breaks down with different kinds and levels of work and varying stages of the relationship. These breakdowns are manifested in problems relating to various themes such as knowledge transfer and identity. In the Sierra case, we discussed a number of communication issues relating to transferring knowledge such as different accents and difficulties in expressing concrete information about encultured and tacit knowledge. These issues are also illustrated in the Japanese case in chapter 9. Processes of knowledge transfer and communication are intimately linked to place-based and contextualized understandings that cannot be developed through the use of ICTs. These insights emphasize the need to distinguish carefully between knowledge that is space- or place-dependent, and to develop different strategies to deal with their transfer.

Contributions to social theory

We first outline in box 10.6 various features of the 'model of' and 'model for' relationship between space–place and globalization.

> **Box 10.6 'Model of' and 'model for' relationship: space–place and globalization**
>
> - Promised growth and cost saving through distance work is in conflict with the proximity needs of managers which come with such work; these processes are inherently contradictory and self-destructive
> - Working in space requires large-scale standardization but these standards cannot work without having a 'place'-like understanding; this interplay leads to redefinition of the standards in play
> - Telecommunication links help to create spaces for distance work; increasing bandwidth enables more interdependent work potentially to take place which, however, raises the need for a more 'place' kind of understanding

The ability to carry out distance work, universality and the associated economic growth are the underlying arguments for globalization, which GSAs serve as a 'model of'. Under such conditions, firms assume that they can have software developed in India or China or Russia using similar business strategies and large-scale standardization. For such standardization efforts to work in practice, however, a 'place'-like understanding is presupposed. Managers express varying reasons for proximity: gaining a comfort level or increased visibility, tighter control and the development of tacit understanding are all needs for proximity. The tensions that arise lead to changes in the nature of work outsourced, redefinitions in the meaning of autonomy and control and shifts in the patterns of communication. These micro-level dynamics in relation to space and place help to develop insights into macro-level social theory.

Castells (1996) emphasizes the importance of the 'power of flows' over the 'flows of power' in the network society. However, the 'space and place' metaphor, when integrated with the material and geographical realities, helps us further to understand how the relationship between power and flows rolls out in practice. Beck (2000) describes the 'despatialization of the social' as a key contour of the 'brave new world of work'. This is characterized by the increasing interdependence between nations, growing importance of transnational institutions and actors and the growth of multi-cultural identities. Beck argues that, as contrasted with the earlier 'simple globalization', in current conditions of reflexive modernization relations are changing not only externally between transnational actors but also *internally*. What goes to make up the 'organization' or its culture or politics becomes inherently questionable because of the breakdown of the principle of territoriality that guided the past. The 'space and place' analytical focus helps to clarify some of these processes that enable the 'deterritorialization of the social': understanding is aided by examining not only the simultaneous occurrence of events between organizations in different countries linked by a GSA, but also those in the same place. Not only are the remote and alien coming together in a GSA but also, as Appadurai (1996) notes, there is an increased *cultural distance* between fellow citizens and neighbours. In an Indian firm providing GSW services, there will be employees working with Japanese or North American clients; these clients have very different work patterns (time–space conditions), security conditions that restrict mutual interaction and a separate technological and market focus. These different groups to a large extent function like 'islands', distant from each other, and in many cases find a greater sense of identification with their alliance partners in the foreign land than with their fellow employees located physically next door.

In outlining this new world of work, Beck argues that 'capital is global, work is local' (2000: 27) and economic processes lose their fixed spatial attachment, as was the case in the industrial society:

Geographical distance thus loses much of its significance as a 'natural' limit to competition between different production sites. In the 'distanceless' space of computer technology, every location in the world now potentially competes with all others for scarce capital investment and chief supplies of labour. (2000: 27)

A dialectical perspective of space–place makes us question such a generalization. Our case studies have emphasized that even in work situations such as GSW, the need for proximity and place cannot be eliminated and geographical distance reaches its 'natural' limits with certain kinds of work and stages of the relationship. The growth in the phenomenon of 'nearsourcing' discussed in chapter 1 also emphasizes the economic argument made by firms which proposes software development in conditions of 'nearness' rather than 'remoteness'. Firms in Mexico and the Caribbean are thus basing their management strategy on emphasizing their geographical 'nearness' to the USA. Firms in India, however, are stressing lower costs owing to distance. These differences demonstrate that economic arguments cannot be formulated based only on cost of production and labour but need to include a number of other issues such as management overheads for dealing with distance and the direct marketing and opportunity costs of operating remotely from potential end-users. The use of 'space and place' as metaphors to understand globalization is incomplete and needs to be integrated with the material realities of geography, history, politics and economics.

Complexity of knowledge transfer

Key features

In box 10.7 we first summarize the key features relating to knowledge transfer.

Box 10.7 Key features of knowledge transfer

- GSW can be conceptualized as reflecting 'knowledge work', including knowledge about products, processes and practices that is being articulated, transferred and interpreted by actors within a 'community of practice'
- Knowledge issues shape the GSA process issues, including the onshore–offshore mix of developers
- Reflexivity of knowledge and the processes of its intensification
- Dependence of knowledge workers
- Knowledge issues need to be analysed within a temporal dimension

---

Through the discussion of knowledge transfer in the Sierra case, and also from the other cases, we extend the critique of the view that knowledge can be treated as a 'commodity' that can be codified, formalized and seamlessly be transferred across time and space. The extension has come by way of discussing knowledge transfer issues in the context of GSW where additional dimensions of time, space and cultures need to be taken into account. These dimensions raise additional complexities owing to the *tacit nature* of certain forms of knowledge and its embeddedness in the practice of actors carrying out their everyday work. These complexities have been discussed at length in the Sierra case, and also emphasized in the Japanese case in chapter 9 with the discussion on 'professional' and 'organizational' knowledge and its implications for shaping communication practices. The temporal perspective emphasizes how varying needs for knowledge emerge at different stages of the project, and how different kinds of complexity arise as a result. We have argued that some of these complexities can be better understood through the 'community of practice' perspective that focuses on the everyday practice of software developers and the contextual influences which shape them. The 'community of practice' perspective needs to be integrated, for example, with the issues of power and control that shape the context within which these practices are structured.

One important aspect of dealing with complexity has been the quest of the firms involved to develop an appropriate mix between *on-site and offshore work*. While placing developers on-site adds to cost, it provides the potential to deal with some of the complexities of knowledge transfer. This mix varies with the temporal stage of the project, the kind of technology work that is being carried out and the professional/personal experiences of the people involved. These issues shape how much of a shared understanding exists between the two sides on what needs to be done and how, and the potential of ICTs to bridge the gaps in understanding. Balancing these two issues of cost and knowledge is a key challenge, especially in understanding how much work can be moved offshore and under what conditions. There are ongoing tensions and negotiations during this process of finding a balance, with significant implications for shaping the GSA process.

The conduct of knowledge work is fundamentally dependent on the capability and stability of knowledge workers. These issues have been raised in the cases such as GlobTel, who throughout the relationship were struggling with the issue of attrition. The Sierra case demonstrated the problems that a small firm has in attracting and retaining high-quality staff in the light of competition from the large and more glamorous MNCs. The case analyses also show not only the reflexivity of knowledge but also the intensified speed through which it occurs. This aspect is seen in Sierra's move to e-commerce or GlobTel's 'right-angle turn', as well as ComSoft's move into the telecommunications domain as compared to their previous EDA focus. What is most interesting in all these cases is the speed and suddenness with which the change took place, and the significant implications this had for various process at multiple levels of the relationship.

Inter-case and inter-theme comparison

Issues of knowledge transfer are fundamentally implicated in all aspects of GSW, including standardization, identity, communication, space/place and power and control. Standardization of knowledge systems takes place in many ways, such as through the use of certified software development methodologies, development of universal 'templates' for management and the implementation of various forms of global 'best practices'. Implementation of these standards is never unproblematic and the resistance that develops as a result of any attempts is key to understanding the evolution of the GSA process. Standardization of knowledge systems is inextricably linked up with issues of power and control, as it reflects whose standards are imposed upon whom. In the Gowing case, for example, Eron developers were made to use specified structured methodologies for software development; this requirement served as an instrument to bring control and discipline within the Gowing structure.

Knowledge issues are fundamentally implicated in questions of *identity* since developers, as discussed in the ComSoft case, are seen to find meaning and relevance in the kind of technologies with which they work. As the GlobTel case demonstrates, working with legacy systems has negative implications on the self-worth of developers, often leading to their flight from the organization. The temporal perspective on GSAs also emphasizes how the relation between knowledge and identity evolves and is redefined over time. At the early stage of the relationship, when the developers are new to the technology domain, they do not mind working with the more routine tasks of bug-fixing and testing. However, with time as the developers increasingly gain knowledge their expectations also change; they expect higher-value work and better conditions. As the GlobTel case demonstrates, if these changing expectations are not met, frustration and resistance develops. This relation between nature of work and expectations is not determinate and industry developments can add new opportunities and challenges. This is demonstrated with GlobTel's 'right-angle turn' when the advent of the Internet redefined the focus from the need for knowledge about GlobTel's telecommunication products to an emphasis on data and the Internet. The Indians believed that this shift in knowledge focus placed them on a 'level playing field' with GlobTel, as it reduced some of the earlier knowledge differentials. However, the advantage of this 'level playing field' was offset by the fact that the Indians were considered to be too remote, in 'space', from the end-users of these new technologies which were primarily located in North America and Europe. Proximity to the end-users and the need for a 'place-like' understanding was seen a basic precondition for the development of new Internet-based technologies.

Contributions to social theory

In box 10.8 we first summarize some key features of the 'model of' and 'model for' relationship between knowledge transfer and globalization before discussing specific contributions to macro-theory on globalization.

Box 10.8 'Model of' and 'model for' relationship: knowledge transfer and globalization

- Global opportunities for knowledge work enable movement of developers, which in turn leads to variability of knowledge quality, with project-level implications
- Despite the attempts to standardize knowledge systems at one end, there is still a strong dependence on individuals
- While technologies such as videoconference provide the potential for disembedded knowledge systems globally, they depend on the local situation to make them work
- Prior experience of work in global contexts (UK and USA) shapes norms of work in particular conditions and through work in different contexts new norms are developed
- Ability to be effective in the global marketplace is still dependent on the effectiveness of local networks – to be able to recruit from quality educational institutions, for example
- 'Born global' firms are at a disadvantage in the light of the brand name and infrastructure of large MNCs
- Applying universal criteria to define forms of knowledge varies with cultural contexts
- Globalization intensifies processes of reflexivity

Knowledge, knowledge work and knowledge workers are key themes surrounding contemporary discussions of globalization and the information society, as well as the digital divide. These discussions are visible in different domains including business organizations as 'knowledge management' and international development as the 'digital divide'. The global importance of this issue of knowledge can be gauged from the fact that the World Bank Annual Report (1998) was titled *Knowledge for Development*. The basic assumption made in contemporary debates is that the commodity 'knowledge' can be codified, formalized and torn away from its local context and rearticulated across other time, space and cultural domains. This process of *knowledge rearticulation* is emphasized by Giddens (1990) through his concepts of 'disembedding mechanisms' and 'time–space distanciation'. Giddens argues that these are key dynamics of contemporary globalization: for example, the assumption underlying distance education initiatives is that the knowledge exchanged between the teacher and student in the setting of a classroom can be formalized as commodities representing 'course materials' and through ICTs distributed to students all over the world.

The debates around the 'digital divide' primarily concern the issue of inequity in access to knowledge, and the question of whose knowledge is being transferred to whom and the appropriateness of it. While we have not directly touched upon these issues, we have discussed a related issue of intellectual property (IP) that raises some similar concerns. For the Indian firms to continue to move up the value chain, they believe that they should in the longer run gain some IP ownership. They thus seek to change the pricing models from one based on 'time and material' in which they get paid for their *labour*, to one based on royalties or licenses where they are compensated for the *knowledge and value* they bring into the process. However, as we have seen, this ambition is rarely realized. The knowledge differential is difficult to bridge and ownership continues to remain with the sourcing company. The Indian firms often feel resentment and seek alternative opportunities in the global marketplace, such as in Japan.

Although attempts to standardize and universalize knowledge systems are a continuous quest of the sourcing company, limits are reached owing to particular local conditions. Videoconferencing provides the potential to disembed knowledge systems; however, much to the dismay of the Sierra management, it first found that it could not easily clear the equipment through Indian customs and later saw that power fluctuations negated the effectiveness of the videoconference meetings between staff in international locations. The original aim of Sierra was to have knowledge about development taking place in India transparent to all, including clients. However, because of the physical distance of the end-clients in the UK from the Sierra premises, the end-users could not easily participate in the videoconference sessions and therefore the aim of 'opening up the black box of knowledge' could not be fully realized. We also see how the attempt to universalize knowledge systems is paradoxically highly dependent on individuals like Mitra in Sierra (chapter 7), who did not seem to have the global vision that he was responsible for implementing. These examples demonstrate the way in which the global depends on the local. The converse is also true: small 'born global' firms like Sierra are at a disadvantage when competing against the global brand names and financial power of large MNCs like GlobTel. Analysing these linkages between the local and global can provide valuable insights into various aspects of social theory, such as how disembedding occurs or how the dynamics of the 'space of flows' rolls out in practice.

Reflexivity of knowledge systems is another key feature of globalization (Beck, Giddens and Lash 1994). We have discussed various examples of how these needs for reflexivity arose from different global conditions and how firms responded to them at a micro level. Sierra, in response to the trends towards e-commerce, decided to shut down its India operations and move the developers to other Western countries. GlobTel, in response to the challenges raised by the Internet, decided to take a 'right-angle turn', which had significant implications for their India relationship. ComSoft, responding to what it saw as declining business opportunities in North America, decided to focus primarily on Japan in the telecommunications domain. The processes of reflexivity and how firms and individuals respond, raise issues of unintended or side effects. GlobTel could not have envisaged that with the learning they provided the Indian firms would explore new avenues even at the cost of their relationship.

Power and control

Key features

We first present in box 10.9, some key features of power and control.

Box 10.9 Key features of power and control

- A relational perspective to power is emphasized through links between power, culture and knowledge
- Cultural norms help to define who can exercise power and how, and how much is acceptable
- Another aspect of the relational perspective is the power–knowledge linkage, which helps to shape mechanisms for 'control at a distance'
- Mechanisms for control at a distance include the making of 'drilled individuals' drawing upon regimes of truth encapsulated in 'non-human actors' such as software development methodologies and ICT-enabled project monitoring practices
- The power–knowledge continuum is never static and varies with temporal stages of the relationship
- Structures of power are a function of history and geography and thus extremely difficult to transcend, even through the use of ICTs
- The 'informating capacity' of new ICTs introduces new dynamics of power, for example through the ability to 'micro-manage' project progress

---

We have emphasized a relational perspective to issues of power and control. These relationships are mediated through aspects of knowledge and culture; they play out in different domains. The power–culture relationship establishes who has the power to define norms of work, to what extent and for whom. Mitra was frustrated because of the norms of hierarchy that he felt existed among the Indian developers. He believed that a creative process should involve all people speaking freely; in contrast, the Indians wanted him as the 'boss' to define the norms for them. The Japanese case in chapter 9 demonstrates very different norms at work where decisions are made by large-sized teams engaging in prolonged negotiations. The Indians, because of their prior North American experience, preferred a relatively individualized form of hierarchy and working; they were at times frustrated by the Japanese approach.

The power–knowledge relationship is another key element of the relational perspective that helps to shape the processes of 'control at a distance'. Individuals can also serve as instruments of their own control. They are 'drilled' through the use of standardized knowledge encapsulated in certified software development methodologies, well-specified and structured project management practices that are monitored through the use of various forms of ICTs (such as videoconferencing). ICTs, with the capacity to make visible action at a distance, can potentially also serve as instruments of control. Elaborately defined project management practices also serve to structure the work at a micro level and further 'drill' the individuals. So while ICTs provide a number of interesting possibilities for instantaneous and simultaneous coordination of software development activities across time and space, these possibilities are structured by the socio-political context of their use.

The use of these new ICTs, software development methodologies and project management practices are a function of history and geography. They can never, therefore, absolutely transcend or break down the power and control structures. Historically it has been the MNCs that have had the financial muscle and brand name to engage in global work. Over time, they have built up brand names and images that are difficult to compete with, especially for the smaller firms like Sierra. Sierra experienced this handicap while trying to recruit high-quality staff: it found that the developers preferred to work with the larger firms. The historical/colonial relationship that existed between the UK and India can still be seen in play: it is not a coincidence that it was Gowing UK, which attempted to 'neutralize' Indian accents rather than the other way around. Again, for videoconferencing to work, the meeting times of the two sides need to be synchronized; in the GlobTel case Indian developers resented that it was always 'their time' that was being compromised to accommodate the 'GlobTel time'. Meetings were typically scheduled at times that were late at night for the Indians so that the North Americans could walk in fresh with a cup of coffee while the tired Indians were waiting to go home. This example illustrates that geography cannot be eliminated by ICTs, but in fact needs to take into account the structures of control and power.

Inter-case and inter-theme comparison

Issues of power and control are linked with a number of other themes. As discussed in relation to standardization, power and control issues are fundamental as they concern who has the ability to impose what standards over whom. Standardized methodologies for software development and project management are themselves encapsulations of the power–knowledge relationship. However, the dialectic of power and control is always in operation, since there is inevitably a resistance and backlash to imposition of standards, and through negotiations standards will be revised. Communication involves the development of norms around the choice of technology used, the frequency of use and even the nature of the content of the messages. Issues of power and control shape whose norms are put in place. Even despite the Indians feeling excessively micro-managed, GlobTel decided to have weekly project-progress monitoring meetings on Monday mornings, and set up a structure that included minutes for the meeting agenda being exchanged on Friday before the meeting.

The metaphors of space and place, we have argued, need to be integrated into the material realities of economics, geography and politics. Power and control issues are fundamental to this integration. As in the MCI case, we described the negotiations that took place as MCI proposed establishing a proximity-development centre to develop a more place-based understanding of GlobTel's practices. However, GlobTel shot down the proposal on economic considerations. In contrast, when GlobTel proposed setting up its own offshore development centre in India, despite Indian resistance, the centre was set up a few years later. Questions of identity are also linked to issues of power and control. As discussed earlier, GlobTel's global manager initiative could at one level be seen as an exercise of power in defining its frame of reference for the construction of identity. Knowledge and power issues have been discussed extensively, especially as they relate to the development of mechanisms to facilitate control at a distance.

Contributions to social theory

We first summarize in box 10.10 the key features of the 'model of' and 'model for' relationship, and infer from it some key contributions to social theory.

Box 10.10 'Model of' and 'model for' relationship: power and control and globalization

- The assumption of 'frictionless' work in globalization is challenged by structures of power and control; through practice, new notions of GSW are constructed
- Software development methodologies reflect universal knowledge systems on the one hand, and yet on the other are linked to local power structures; through the application and use of these methodologies, power relationships can be reconfigured
- Changes in locally situated power structures can be seen as manifestations of the 'reverse effects of globalization'
- The micro-level dynamics of the GSA process help to shape the relationship between 'flows of power' and the 'power of flows'
- Division and commodification of labour can be an outcome of GSAs, but need to be differentiated with respect to the different kinds of work being done
- GSW, while reducing some historical structures of power and control, also opens up new ones
- While 'control at a distance' is a basic precondition for GSW, the mechanisms for control need proximity to be effective

---

The underlying assumptions of globalization that 'geography is history' and 'distance does not matter' are seriously challenged when we examine the micro-level dynamics of GSW and how they unfold over time. If, as we have argued, structures of power and control are a product of historical and geographical circumstances that can never be absolutely transcended, then the interesting questions are which aspects of history and geography can be transcended, how and under what conditions. It has been argued that in the 'network society' those firms which will reign supreme are those that fundamentally rely on knowledge, and have the capability to *leverage this knowledge* by strategically situating themselves in global informational networks. However, this is not unproblematic. It is still primarily the large firms that have the power to lobby with governments and hence the capability to make large investments in exclusive high-speed networks. In contrast, the smaller firms often need to rely on using networks and infrastructure that may be controlled and regulated by the state. Although in many countries these restrictions are being removed, it is safe to say that this will not be the case everywhere. Smaller firms are thus subject to greater delays in obtaining access to markets and find it relatively more difficult than larger firms to reach economies of scale. However, it is not the case that such handicaps are inevitable. Examples abound of small firms such as Hotmail which established themselves through the strategic development of extremely innovative products and alliances with larger firms. Such a nuanced understanding of how knowledge-based firms operate can be developed through the analysis of power and control issues. This can support efforts to gain deeper understanding of the dynamics in Castells' 'network society'.

GSW, while providing the potential to break down to some extent historically created structures of power and control, can also help to open up new ones. Beck (2000) discusses relocation of work from Germany to parts of the world that MNCs find more cost effective for their operations. This reflects a breakdown of structures of bureaucracy: the bureaucracy in India has historically tried to control the entry of MNCs but the clustering of large MNCs in a city such as Bangalore creates other kinds of power structures, such as the development of an elite group of highly paid IT professionals in the city. This has broader societal implications for property prices and the cost of living, with adverse effects on people employed outside the IT sector. The German 'green-card' initiative has had implications for policies that have historically tried to restrict immigration. Giddens discusses such effects through the concept of 'reverse globalization' – how not only local events are being shaped by global happenings but also the reverse: relocation of jobs to India may lead to issues of underemployment and unemployment in Philadelphia, USA. The 'networked' perspective that we have taken on GSAs helps us to understand the micro-level dynamics of how power- and control-related 'reverse globalization' effects can take place in practice.

The anti-globalization debate emphasizes the kind of dangers and risks that people perceive to exist with GSW arrangements. GSW fundamentally involves the breaking down of software development into discrete modules and then spreading it around the world where it can be done most efficiently. These different modules of development can be micro-managed through the use of ICTs and productivity based on 'time and material' can be measured and rewarded or punished according to the outputs. These characteristics of GSW represent Mowshowitz's (1994) vision of the future of work that involve the 'switching of resources' and 'separation of the means and ends' of production.

Switching of resources implies that global resources from different locations can be drawn upon and switched, implying the means and ends of software development work can clearly be separated from each other. Various authors have criticized such visions in the past. Braverman (1974) and the labour theorists have seen such ways of organizing work as a systematic progression in the development of mechanisms to control the work force. Ritzer (1996) has similarly critiqued the 'McDonaldization' of society through the increasing tendency towards scientific management or 'Taylorization' of work systems. In the same vein, Walsham (1994) has critiqued Mowshowitz's vision, arguing that it will lead to a dehumanization of the work force. While these critiques have to be seriously considered, we argue that a more nuanced vision of the effects of GSW needs to be taken. GSW includes a broad spectrum of task from running call centres in which Taylorist principles can be applied in near totality to higher-level new product design tasks where it may be counterproductive to apply them. In between, there are numerous shades of work amenable to Taylorist organization to a varying extent. It thus becomes important not to make sweeping generalizations about GSW, in either positive or negative terms, but instead to take a case-by-case approach.

Communication

Key features

We first outline in box 10.11 some key features of communication.

> **Box 10.11 Key features of communication**
>
> - Communication in GSAs is shaped by the stereotypes that actors have of other cultures and nations
> - The effectiveness (or not) of communication shapes various aspects of GSA, including the business model adopted to project management practices and to the development of social relationships
> - Effectiveness of communication is a key aspect to be considered when defining the mix of people to be on-site versus offshore
> - ICTs are necessary to enable communication in GSAs, but not sufficient to ensure efficient collaboration
> - Communication depends on various elements including understanding of language and knowledge of the meanings of words and phrases as they are used in different situations and contexts
> - Groups and cultures favour different styles of communication; 'organizational' or 'professional' approaches to knowledge involve communication styles that are 'low'- and 'high'-context, respectively

Various authors (for example, Couch 1989) have described communication to be fundamental in the shaping of social life. In GSAs, the significance and centrality of communication process is magnified since a large part of GSW takes place in conditions of separation where work- and non-work-related interaction occurs through the use of electronically mediated communication. Communication issues arising from the separation of time, space and cultures, coupled with complexities associated with the use of ICTs, help to shape various decisions related to the selection of the business model, the structuring of project management practices and the shaping of social relationships.

These complexities vary with different stages of the relationship as various norms are established, reflected and also redefined in and through ongoing communication. For example, an important norm concerns the amount of contextual information that is provided in messages. The Indian–Japanese analysis brings to the fore how approaches to communication vary with the extent of contextual information that actors reflect and also expect in their messages (verbal or written) that they transmit. Difference in expectations on these issues potentially can lead to *communication dissonances.*

It is often argued that analysing cross-cultural communication in terms of national stereotypes is limited, as it is static and tends to reduce the human agent to the status of a 'cultural dope' (Garfinkel 1967). However, a key implication of our analysis is that these national stereotypes cannot be dismissed since actors draw upon such conceptions in the course of developing communicative action. These stereotypes can in Giddens' (1984) terms be described as 'resources' that actors draw upon in their everyday act of communication. For example, an Indian who sees Japanese in general to be ritualistic and taking a long time in making decisions writes few and short email messages and tries to address issues in face-to-face meetings to the maximum extent possible. However, these stereotypes are never static and are redefined or reinforced in and through the process of communication itself.

Inter-case and inter-theme comparisons

Communication is fundamentally implicated in other themes including space and place, standardization, identity, knowledge transfer and power and control. Gowing UK (chapter 8) made use of consultants to 'neutralize' the Indian accents so that the Indians could be understood more easily by the British personnel over the electronic medium. This reflects Gowing's power over the Indians rather than the other way around. GlobTel through their 'global manager' initiative tried to standardize communication competencies, such as who should discuss what with whom. We have in the discussion on standardization emphasized a hierarchy of socio-political standards in addition to technical standards and their relation to different levels of GSW. As work becomes higher-value (for example, involving design activities), the need for communication increases as compared to lower-level tasks of bug-fixing and maintenance that are less interdependent. Standardization of communication processes can thus be associated with higher-value work that requires a greater degree of interdependent activities.

Complexities of knowledge transfer are centrally associated with issues of communication. A key implication in this regard concerns the relation between approaches to knowledge and communication styles. Through the Japanese case in chapter 9 we analysed how different styles of communication (high- versus low-context) are related to varying approaches to understanding knowledge ('organizational' versus 'professional'). Varying styles of communication cause complexities in knowledge transfer. The Sierra case demonstrated the difficulties in transferring and interpreting information about requirements analysis; such problems arise from issues of tacit knowledge and the lack of contextual understanding the Indians possessed of the domain about which requirements were being communicated. The greater the emphasis on knowledge that is 'place'- or context-dependent, the higher is the complexity in making communication effective over distance, mediated by ICTs in 'space'. The linkage between communication and space and place also has implications for identity. An Indian programmer in Sierra told us that he used two accents, one a British accent to speak to the UK clients (in space) and the other, a 'South-Indian' accent in his home (in place). Switching between these two accents raises broader existential questions relating to the individuals' sense of belonging.

Contributions to social theory

In box 10.12 below we outline key features of the 'model of' and 'model for' relationship between communication and globalization.

Box 10.12 'Model of' and 'model for' relationship: communication and globalization

- The national and cultural stereotypes which actors have are key in shaping communication practices
- The effectiveness (or not) of these processes helps to redefine our understanding of other cultures; this leads to new situated meanings of globalization
- Working at a distance, a basic condition of globalization, meets its limits in the effectiveness or not of communication; this leads to revised understandings of the meaning of what it is to work at a distance, and with it the meaning of globalization
- The assumption that ICT-enabled communication can 'make geography history' is problematic and actors realize the need to understand both geography and place in order to communicate effectively
- The effectiveness of communication shapes decisions relating to what work can or cannot be outsourced, and with it the opportunities one has or does not have in the global marketplace

---

The focus on understanding communication processes in the GSW task helps to develop a number of contributions towards social theory, including the role of national stereotypes in shaping global work. We have argued that national stereotypes should not be dismissed on academic criteria as being inadequate, but should instead be treated seriously as resources that exist as memory traces in actors' heads and which are drawn upon in shaping communication. In this way, we extend structuration theory by incorporating a micro-level understanding of the communication process and also incorporate in it the important *cross-cultural* dimension.

Giddens' theories of modernity and globalization emphasize the role of disembedding and re-embedding mechanisms in shaping contemporary life. While this idea sensitizes us to the macro-level dynamics that are in play in globalization, it is limited in explaining how these dynamics roll out in practice. A focus on communication processes emphasizes the reasons why complexities in knowledge transfer occur. An analysis of the manner in which actors take one communication style and vocabulary from one cultural context to another also helps to understand how 'hybridized' models of globalization are created. In Japan, the Indians start with a dominant North American frame of reference, but over time start to introduce a Japanese flavour into their interactions.

We can also add to our understanding of the dynamics of Castells' (1996) 'space of flows'. Castells discusses aspects of *simultaneity and synchronicity* of new ICTs and their potential to shape the flows of knowledge in space. However, situating the use of these ICTs for communication within a broader cross-cultural context helps us to understand what some of the inherent limits in the use of these technologies are.

In summary, we have attempted to develop a theoretical synthesis of the various themes that have been discussed in the book. These themes have been discussed in relation to their key features, their relation to other themes and the contributions that can be developed for social theory. The next section considers future extensions and limitations.

10.4 Extensions and limitations of the research programme

A key aspect of our multi-case analysis has been the emphasis on process and context and the linkages that exist between the two in relation to GSA relationships. *Process* refers to the manner in which the relationship evolves over time, specifically how it is initiated and the explorations that go on around it, its growth and the dynamics of how it stabilizes or breaks down. This process is examined in terms of the changes in work, in the quantum and type of technologies with which people engage, the changes in numbers of people deployed in projects, the associated growth in infrastructure and the maturation of management practices. *Context* refers to the ongoing processes of globalization, how they stabilize, strengthen or destabilize GSA relationships and how meanings of globalization change as a result of these micro-level dynamics. The dynamics of globalization that we have analysed through the metaphors of 'out of control', 'networks', and 'risks' should be taken into account, as changes in technological trends in the industry, redefinitions of global markets and changes in government policies as a result of global opportunities and challenges have significant multi-level impacts for the GSA relation. *Linkages* between context and process have been studied through the six conceptual themes that were discussed in greater depth earlier in this chapter. This focus on the *mutual linkages between context and process* is a response to Walsham's (1993) critique of IS research with too much focus on content, and the neglect of process, context and their linkages. This approach also responds to Walsham's subsequent arguments (1998, 2001) about the need to examine how micro-level analysis of IT-enabled social transformations contributes to macro social theory.

A perennial question confronting IS research, especially in the interpretive tradition, is about the generalization of research findings. The intention, as is the case of researchers in the interpretive tradition within which this book is grounded, is not to try to develop predictability of results but to come up with 'thick' descriptions of a phenomenon that provide deep and useful insights into the nature of the GSA process and how it evolves over time. These descriptions should be coherent, believable and based on an interpretive process that is clearly explicated and understandable. As a result of our interpretative process we have not developed 'statistical generalizations' that aim to predict but 'analytical generalizations' that help to further our understanding. These analytical generalizations can be seen at two levels:

- At a more meta level, the 'model of' and 'model for' relationship proposed and elaborated in the book helps to provide a conceptual frame in which GSAs can be examined more generally. Such a conceptual frame, we believe, can be also useful for analysing other IS-related phenomena such as knowledge management, infrastructures or implementation.

- At a more micro level, we have identified and elaborated six concepts that help the analysis of GSAs within the meta-level framework.

While we are not going to say that the dynamics of standards will unfold in any predicable fashion, we make more abstract generalizations that there will be a hierarchy of standards in operation or that attempts to implement standards will be subject to ongoing negotiation and redefinitions.

Our research, like all studies, has its limitations:

- In the research design we have studied only India as a country with which firms from other countries have GSA links. As discussed in chapter 1, there is now a whole range of countries, such as China, Russia, the Philippines, etc. that are providing competitive GSA services. While a focus on one country helps to provide focus, it is limited in developing more general implications applicable to other countries.

- This limitation we are trying to remedy in the next phase of research (started in 2001) in which we are also studying firms from Eastern Europe and Russia. Another extension being carried out in this new phase of research is to look at different sourcing countries. Specifically, we are focusing on Norway and Germany and initial interviews have already commenced.

- Our study has had a primarily macro-level focus on the strategic issues relating to the management of the relationship. Through our interviews we were generally trying to understand the challenges that firms and individuals were experiencing, how these were changing over time and what the responses were to them. As a result of this focus, we did not have very detailed project-level information, nor did we follow a particular project over time. This is a significant gap in our analysis because the micro-level dynamics of the projects both shape and are shaped by the GSA-level issues that are situated within the broader globalization context. To follow these project-level dynamics we would need access to data such as email communications, study the software code and examine the requirements documentation. Access to such information is difficult to obtain. However, in one of our new case studies of a Norwegian–Indian relationship we have managed to get such access and we have been invited to sit in on the meetings of the developers and clients where the requirements are developed in Norway. We will then be also able to examine the transfer of these requirement documents and interview staff at both ends to see how the requirements were then interpreted. Such an analysis, we believe, will strengthen our focus further in due course.

- Although we have discussed a set of themes in the cases, some important ones have not been examined. We have not looked at financial data such as income statements and balance sheets. Exclusion of financial issues is a serious limitation that we will address in the next phase of the research. Trust, another issue that is currently an important topic for discussion in the IS community, has also not explicitly been addressed. While we have made implicit or tangential references to the issue of trust themes which in different ways help to develop insights into the concept, we have not made a full-blown analysis of the theme along the lines of the other six. However, we plan to examine the issue of trust more completely in the new phase of research.

A criticism often made of IS researchers concerns 'relevance', or rather the lack of it, in their studies on practice. We believe that theory and practice cannot be separated and that there exists between them a *mutual and inseparable* relationship. We develop the managerial implications from our study in chapter 11.

BIBLIOGRAPHY

Appadurai, A. (1996). *Modernity at Large: Cultural Dimensions of Globalization*, Minneapolis and London: University of Minnesota Press

Beck, U. (1992). *Risk Society: Towards a New Modernity*, London: Sage
    (2000). *The Brave New World of Work*, Cambridge: Polity Press

Beck, U., Giddens, A. and Lash, S. (eds.) (1994). *Reflexive Modernization: Politics, Tradition and Aesthetics in the Modern Social Order*, Cambridge: Polity Press

Blackler, F. (1995). Knowledge, knowledge work and organizations: an overview and Interpretation, *Organization Studies*, 16, 6, 1047–75

Braverman, H. (1974). *Labor and Monopoly Capital: The Degradation of Work in the Twentieth Century*, New York: Monthly Review Press

Castells, M. (1996). *The Rise of the Network Society*, Oxford: Blackwell
    (2001). Globalization and identity in the network society, *Prometheus*, March, 4–18

Couch, C. J. (1989). *Social Processes and Relationships*, Dix Hills, NY: General Hall

Garfinkel, H. (1967). *Studies in Ethnomethodology*, Englewood Cliffs, NJ: Prentice-Hall

Giddens, A. (1984). *The Constitution of Society*, Cambridge: Polity Press
    (1990). *The Consequences of Modernity*, Cambridge: Polity Press

Harvey, D. (1996). *Justice, Nature and the Geography of Difference*, Oxford: Blackwell

Hofstede, G. (1980). *Culture's Consequences: International Differences in Work Related Values*, Beverley Hills, CA: Sage

Lave, J. and Wenger, E. (1993). *Situated Learning: Legitimate Peripheral Participation*, New York: Cambridge University Press

Morgan, G. (1986). *Images of Organization*, London: Sage

Mowshowitz, A. (1994). Virtual organizations: a vision of management in the information age, *The Information Society*, 10, 267–88

Nicholson, B. and Sahay, S. (2001). Some political and cultural issues in the globalization of software development: case experience from Britain and India, *Information and Organization*, 11, 1, 25–44

Nicholson, B., Sahay, S. and Krishna, S. (2000). Work practices and local improvisations within global software teams: a case study of a UK subsidiary in India, Proceedings of the *IFIP 9.4 Conference on Socio-Economic Impacts of Computers in Developing Countries*, Cape Town, 24–26 May

Ritzer, G. (1996). *The McDonaldization of Society*, 2nd edn., Thousand Oaks, CA: Pine Forge Press

Sahay, S. and Walsham, G. (1997). Social structure and managerial agency in India, *Organizational Studies*, 8, 3, 417–43

Schultze, U. and Boland, R. J. (2000). Place, space, and knowledge work: a study of outsourced computer systems administrators, *Accounting, Management and Information Technologies*, 10, 187–219

van Maanen, J. (1989). Some notes on the importance of writing in organization studies, Harvard Business School, Research Colloquium, 27–33

Walsham, G. (1993). *Interpreting Information Systems in Organizations*, Chichester: Wiley

(1994). Virtual organisation: an alternative view, *The Information Society*, 10, 289–92

(1998). IT and changing professional identity: micro-studies and macro-theory, *Journal of the American Society for Information Science*, 49, 12, 1081–9

(2001). *Making a World of Difference: IT in a Global Context*, Chichester: Wiley

World Bank (1998). *Knowledge for Development*, World Development Report, Washington, DC: World Bank

11 Managerial implications

11.1 Introduction

This chapter is concerned with developing implications to aid managers currently engaged in establishing, managing and strengthening GSA relationships. Our aim is to develop implications that are grounded in rich empirical data that has been subjected to rigorous theoretically informed analysis of 'how the GSA process evolves over time'. This analysis develops relevant management implications around the six analytical themes discussed so far: standardization, identity, space and place, knowledge transfer, power and control and communication. The theoretical and empirical basis of the analysis helps us to transcend providing mere prescriptions and 'how-to-do-it' guides for GSA management and instead, develop a set of questions around key tensions that managers can analyse in the process of managing GSAs.

The GSA literature has a number of examples of prescriptions for management (see for instance Apte 1990; Apte and Mason 1995; Rajkumar and Mani 2001; Karolak 1998; Carmel 1999). For example, the rule to 'outsource only structured tasks' may work in certain situations but this is not always the case. Such rules, while useful in sensitizing managers to issues of importance for managing GSAs, are limited in the face of the complexity of the unfolding process over time, which makes it difficult to apply 'best practices' or 'universal methodologies'. Prescriptive rule-based approaches tend to downplay the local and contingent situation in favour of broad generalizations and ignore the possibility that what worked well in one relationship may cause problems in another. In the Eron–Gowing case (chapter 8) the perceived deference and compliance of Indian developers was a key element of Gowing's strategy, while the same traits were seen as a major problem in Sierra. While North American clients tend to favour outsourcing of highly structured work, we found the Japanese firms to be more willing to venture into relatively unstructured and new-technology domains. Thus the inter-case comparisons we conducted help to emphasize how *varying contextual conditions* can shape relationships differently.

While technological advances in telecommunications and in the use of various software development methodologies imply the normalization of global sourcing of IT services, effective management strategies are far from being a given. The tremendous complexity of the environment forces us to look at aspects of control and risk associated

Table 11.1 Key implications for GSA management

Metaphor	Implications
Out of control	Complex, volatile nature of GSA environment
Risk	Problem of risk redistribution, assessment, control and management in a situation of change
Networks	Evolving options and tensions can be understood with a network perspective

with GSA relationships quite differently in comparison to traditional businesses. We have tried to develop such a reconceptualization in analysis through the metaphors of the 'runaway world' (Giddens 1990), 'risk society' (Beck 1992) and 'network society' (Castells 1996). These metaphors, while helping to theorize about the influence of globalization processes that cause turbulence and complexity in GSAs, are also extremely useful in developing practical implications. Placing GSAs within the context of globalization has a direct implication in emphasizing the need for managers to acknowledge the extreme complexity and turbulence of the context in which they operate and that changes and surprises in the relationship are the norm rather than the exception. As a result, managers have first and foremost to acknowledge the 'out-of-control' nature of their task and the futility of attempting to build tightly controlled and predictable environments. We summarize in table 11.1 some key implications for management that arise from the three metaphors discussed above.

This 'out-of-control' aspect is emphasized by Hutton and Giddens (2001) who describe contemporary life as being 'on the edge'. Macro-level changes as a result of interconnected economic, political and other global processes have direct implications for firms and individuals engaged in the GSA domain. Examples abound of these macro-level changes and their impact on GSAs. The significant downturn of the global economy and the IT industry in particular in 2001 led to the 'benching' of large numbers of programmers in India. The entry of China into the WTO in 2001 can only stimulate new avenues for outsourcing. Faced with this new competition and the need 'not to put all their eggs in the US basket', supplier firms in India and elsewhere have actively begun to explore new European markets such as Germany, Italy and Scandinavia, while also trying to establish their own software development centres in China and Vietnam where labour costs are relatively cheaper than India. The devastating 11 September terrorist attacks on New York have introduced new risks with the tightening up of immigration policies. This will directly impact on GSA strategies that depend on the free movement of developers across global borders. The extremely fragile political and military situation existing between India and Pakistan raises serious doubts among potential and existing GSA customers about continuing or entering into new GSA arrangements in India. The birth of the Euro and the combined efforts of the European Union to deal with IT skills shortages will no doubt become an important factor for developers to consider when they make a choice of whether they should go to North America to find employment. Changes in the technological and business context open up new challenges and also opportunities for GSAs; for example Sierra (described in chapter 7), because of wider global business trends (toward e-commerce), decided that offshore development in India was no longer viable for them. But as a result of this decision the Indian developers were relocated to the other Sierra offices (in London, San Francisco and New York), which no doubt led to other unknown opportunities (of working with new technologies) and risks (of potentially being 'benched' in a downturn).

These rapid environmental changes have direct implications for the networks and business models adopted by firms for GSAs. Offshore software services were once the sweatshop of the IT industry for low-cost programming and data processing. This is no longer the case and current global outsourcing trends include applications service provision (ASP), enterprise resource planning systems (ERP), as well as a range of business process outsourcing (BPO) services. GSA is no longer just concerned with developed countries moving work to developing countries for cost savings but with securing opportunities in a diverse and thriving global services industry. In India, ASP and BPO trends are likely to display a similar value chain trajectory to the software industry. The more mature Indian software outsourcing companies expect to continue to move up the value chain by localizing their services in new countries and markets and also by entering into JV arrangements with foreign consulting firms. India's Mastek, for instance, entered into such an arrangement with Deloitte Consulting in 2001 with a view to expanding Mastek's technological strengths with domain expertise in customer businesses. Existing strategic business models of GSA options need to be reconsidered in the light of the potential for localization of firms in new countries and the establishment of new kinds of business relationships.

In chapter 1, we discussed the influence of global trends, organizational forms and the nature of GSW itself on the progress of a GSA relationship. Planning and implementing strategic directions for GSA is problematic, as environmental turbulence continuously creates the need to make sense of benefits and risks. The benefits of GSAs shown in the GlobTel and Gowing cases lay in flexibility, skills and resource availability. Strategies aimed at obtaining cost savings alone can be disappointing because of the significant hidden management costs. Royal Sun Alliance conducted trials and concluded that the average saving was 30–35 per cent with 10–12 per cent expenses in intangible costs of management time spent on control (*Computer Weekly*, 16–17 May 2001). Agency fees, handover costs, rising salary bills and cost of communications links rapidly eat into expected savings. The future success of GSAs may lie with those companies that do not only have cost saving objectives, but with those who also have other strategic objectives such as new markets or technology development.

A network perspective on GSA permits consideration of various possible relationship forms and multiple suppliers in different countries (Faulkner and Child 1998). The cases show a range of GSA approaches based on a single sourcing (Gowing and Sierra) or multiple sourcing strategy (GlobTel), all involving India. Adopting a network perspective emphasizes that other countries also need to be assessed based on technical specialism, costs, market potential and maturity of the industry. Assessment of risk needs to take into account cost savings, language, distance, infrastructure, available incentives and political stability. These are difficult to reduce to a list of factors that can be quantitatively modelled. The risks faced by GlobTel's GSA are very different to those faced by a small company like Sierra because of the very different scales of operation. Multinationals such as Microsoft are actively courted by states in India (such as Andhra Pradesh) and have the size and financial muscle to overcome some of the contextual variables so troubling to Sierra.

The cases also show a range of relationship types, from the wholly owned subsidiary (Sierra), to 'vendor-contract' relationships (GlobTel), to third-party outsourcing (Gowing). These different organizational forms, as discussed in chapter 1, have varying implications for risks, such as that relating to IP ownership. To try and mediate these risks a range of middlemen and brokers has emerged acting as bridges between onshore and offshore in setting up arrangements or incubating offshore subsidiaries. The 'near-sourcing' option opens up another whole new range of possibilities of outsourcing to firms in Mexico, the Caribbean and even Canada. These different possibilities heighten the need to look at GSAs with a 'networked' perspective and evaluate the possibilities and risks that result.

Key challenges in the management of GSAs can thus be encapsulated in the metaphors of 'out of control', 'risk' and 'networks'. While 'out of control' refers to the futility of implementing tightly controlled operations in such a turbulent evolving environment, 'risks' refer to the multi-level challenges that arise as a result of the new forms of GSAs and associated technologies. The network approach to the ongoing strategic planning of GSA emphasizes the need to make simultaneous sense of the range of evolving business relationships and services available globally.

We now discuss the specific management implications of GSA, not with a view to prescribing 'recipes' for management but to identify particular areas that are dynamic and subject to processual tensions. We formulate five questions around the thematic areas discussed earlier to provide a frame of reference for critical reflection on the key challenges in managing GSA relationships:

- *Managing knowledge*: How can knowledge be managed across time, space and culture in an effective manner?
- *Managing people*: How can managers attract, retain and motivate their workforce in the light of global demand for scarce human resources?
- *Managing communication*: How can communication practices be managed effectively?
- *Managing relationships*: How can relationships be initiated, fostered and sustained over time?

And last but not least:

- *Managing ethically*: How can managers engage in GSAs in an ethical manner?

The remainder of the chapter is organized around these questions separated into our thematic areas. For each question, we present an overview of the thematic area together with the key challenges and issues illustrated with examples. Finally, we present a summary of the questions that managers may ask in relation to unpacking the complexities of specific GSA situations that will help them to develop creative responses.

11.2 Managing knowledge

In various ways, all the cases in this book deal with the issue of *knowledge in GSW*. In the Sierra case, we describe how managing knowledge in a GSA concerns the development, transfer, interpretation and use of knowledge by a 'community of practice'. We have described how knowledge transfer is related to products, processes and practices and can then be further analysed through the lenses of encultured, embedded and encoded knowledge. Related themes of knowledge and power are discussed in the Gowing case. Knowledge was centrally implicated in GlobTel's relationships and reflected in the negotiations around the issue of standardization. Further issues of 'professional' and 'organizational' knowledge and their implications for communication were discussed in the Japanese case in chapter 9.

Transferring knowledge about practices (*encultured knowledge*) is problematic, as it tends to make sense only in local contexts, thus making the instrumental perception of a seamless transfer unrealistic. At Sierra, ICTs (such as videoconferences) were largely unable to bridge the 'community of practice' between the staff in the UK and the India centre, which existed largely as separate entities despite Mitra's best efforts to transplant the standardized UK way of working in India. The Japanese companies were confronted with various knowledge-related issues, because of their different approaches to knowledge as compared to the Indian side. In the Japanese companies, knowledge tended to be created through an experience-based learning described as 'organizational knowledge'. This knowledge tended to rely on tacit, often undocumented, knowledge built up over a long period of time of working in the firm. The Japanese found it difficult to communicate this form of knowledge to the Indian companies which by contrast adopted a 'professional' approach to knowledge which was relatively more explicit, relied less on tacit knowledge and more on documentation and formalized methodology. This led the companies to adopt different business models and associated communication strategies for managing knowledge in the GSA.

In GlobTel's case, the different knowledge focus arose from the different domains of operation. GlobTel, being a telecommunications company, focused on a 'product-based knowledge' relating to telecommunications. In contrast, MCI, like the other Indian firms, relied more on a 'software service knowledge' concerned with software skills and project management capabilities. These different frames of reference raised significant challenges on issues such as attrition. Telecommunications product knowledge typically takes many years to develop as compared to the software service knowledge acquired through short-term courses and experience of few projects. Indian software engineers were thus ready to leave and move on to another project and technology after about a year. GlobTel viewed this problem of attrition as extremely serious, and could not quite understand why the Indians left so quickly especially compared to their own long tenures (sometimes even 10–15 years).

Reflecting on knowledge transfer challenges, the Sierra management in hindsight felt that only highly structured work should be sent offshore, as that would involve minimal communication. This insight supports typical offshore models that favour the outsourcing of relatively structured projects (McFarlan 1995). However, GlobTel moved up the value chain to more sophisticated work being carried out offshore supported by large-scale standardization, mature project management processes, extensive similar experience of running GSAs and significant financial investments. The use of 'straddlers' and onshore–offshore teams containing a mixture of personnel from 'both sides' supported the process of knowledge transfer. While it was possible for a large firm like GlobTel to make significant investments to deal with the challenges of distance, for example in the extensive use of expatriates, a small firm like Sierra could not make the same level of investments.

The Japanese case in chapter 9 shows the importance of improvisations in relation to knowledge transfer. Individuals on both sides improvised extensively in an attempt to reach an understanding on the requirements, such as making modifications in the structure and contents of email messages and in making selective use of ICTs. These improvisations are critical and reflect the motivation structures and flexibility that managers must develop in making knowledge transfer in GSA more effective. To move beyond simple coding, development staff need to work towards developing realistic and shared expectations. The GSA knowledge transfer process should be perceived as one of learning, collaboration and improvisation, as compared to the utopia of seamless knowledge transfer, where specifications can be sent to the 'black box', and code magically received back in time and on budget. Such unrealistic expectations, as emphasized through the case of Sierra, will lead only to frustration and disappointment.

In developing a perspective and implications for knowledge transfer, the questions that reflective managers need to consider are summarized in box 11.1.

Box 11.1 Key questions: managing knowledge

- What types of knowledge is it important to transfer (products, processes and practices; encultured, encoded and embedded knowledge)? Where can shared understanding be achieved and what might be the barriers?
- How feasible and realistic is it to transfer certain aspects of knowledge (given the complexity of work and culture)? What alternative forms of transfer can be developed?
- What methodologies and approaches are relevant and feasible (global standardization, local autonomy, structured approaches) to support transfer processes?
- How can ICTs be used to augment knowledge transfer processes? What are the limits of the ICTs and what other approaches should be used?
- How can a cost effective blend of offshore and on-site staff be developed?

11.3 Managing people

GSW-related knowledge work is made possible only by attracting, retaining and motivating capable knowledge workers. This key challenge of managing people is shaped by the size and context of the relationship and has implications for implementing structural innovations to manage attrition and motivate staff.

Sierra's India centre faced the constant problem of attrition that had magnified consequences compared to a large MNC like GlobTel. The impact of losing a single developer in a 20-person centre is magnified as compared to a similar loss in a 500-person centre. This loss is felt more if large amounts of money have been invested on training, on rotating the staff to onshore customer sites to help develop customer-specific knowledge and close personal working relationships with onshore colleagues. Recruitment and retention of staff will always be a problem for small companies such as Sierra who compete for the best staff with large foreign MNCs offering a structured career path, glamour and the potential for foreign travel and working in more attractive conditions of new technologies and with higher remuneration. Reputation and firm size thus become major factors in attracting the best programmers. Besides these factors, managers need to develop a close understanding of local contextual issues that influence the attraction, motivation and retention of development staff. These complex issues expose companies to risks that are not immediately perceptible or easy to manage. Although the 'streets of Bangalore may be filled with programmers', it is still not easy to attract and retain them, as Sierra found out to their disadvantage. Even large Indian software houses find it difficult to manage the 'out-of-control' nature of global environmental trends such as rapid changes in advanced technological platforms. For example, in 1998 skills requirements were dominated by the North American demand for developers with experience in Powerbuilder. Such work was perceived by many Indian developers as furthering their careers and leading to future foreign travel to the USA. Interviews with Eron developers emphasized their reluctance to work on what they perceived as outdated platforms (such as Ingres), as it would impede their career prospects and opportunities for travel. Similarly, GlobTel in making its 'right-angle turn', decided to give Indian firms primarily the task of maintaining their legacy systems. This decision was expected to potentially cause large-scale attrition of developers working on GlobTel projects. In contrast, some Indian engineers expressed a preference for working in Japan as it provided them with the potential to work with new and cutting-edge technologies being tested in the market.

High attrition rates characterize the Indian software industry, averaging about 30 per cent at the peak in 1998–2000. Although this situation became less intense in 2001 owing to the US recession, it is volatile and subject to change. Most of the turnover tends to be junior developers (1–2 years' experience) who leave primarily to North America for purposes of higher studies or taking up employment. Very rarely is the movement horizontally to other firms in India. Some Indian suppliers have taken drastic measures to curtail the problem by opening software centres in emerging locations such as China and Vietnam where staff attrition is not yet so much of a problem (*Computer Weekly*, 2nd August 2001). The top IT firms in Vietnam claim to be able to keep project teams together for months at a time.

In order to develop effective structures for recruitment, motivation and retention, large Indian software outsourcing companies offer structured career paths. Mastek, a top 20 Indian software house, achieved structured accreditation for the Capability Maturity Model (CMM) and People Capability Maturity Model (PCMM). This is an indicator of well-structured and defined career paths at the highest level (5) (at the time of writing, Mastek was the only company in the world to achieve PCMM Level 5 status). Mastek, like many other Indian software houses, provide share options, interest-free loans and training opportunities as incentives to retain key staff. Trends towards localization by opening centres in the UK, the USA, Japan, Germany and Belgium have meant that Mastek can offer employees overseas experience and mitigate the potential for individuals leaving because of limited opportunities for foreign travel. GlobTel offered Indian staff the chance to come to North America for a year after having served for 3 years, and the possibility for employment into GlobTel after 5 years of experience.

Despite all these efforts Indian software houses accept staff attrition as inevitable and attempt to build networks with external agencies to maintain a steady staff intake. Instead of trying to control attrition, as GlobTel did, MCI tended to focus more on *managing the implications of attrition* more effectively by introducing systems of 'shadowing' and 'buffering' where more developers would be deployed working on the project than were actually billed. Even though this raised salary costs, this was acceptable in comparison to the costs involved in recruiting and training new employees. Many companies try to inculcate a strong sense of local identity to help motivate staff. Mastek, for example, has a reading group where employees come together and read a range of texts for discussion; guest academic speakers are invited to give talks on topics such as globalization and Indian identity. In one such lecture, the academic speaker contextualized India's current position in the global software industry in relation to its historical strengths and leadership in mathematics and sciences. Mastek staff also participate in weekly reading groups on topics as diverse as the writings of J. Krishnamurthy, an influential Indian philosopher. Software houses also try to develop the tradition of 'annual getaways'. Mastek's 'run time' event brings together clients and staff in an environment away from the corporate office to help develop solidarity. ComSoft consciously tries to create a 'family' environment that emphasizes 'Indian' values to provide conditions in which the staff feel that they can do work on a par with the best in the world while being based in India.

Standardization of motivation structures can be problematic. Sierra tried to import motivation structures directly from the UK, including reward and recognition systems. A small firm, with limited experience of operating in India, it had problems understanding what would motivate the Indian team. Mitra, the British manager in India, once told us in exasperation that his American textbooks on motivation didn't seem to work in India. Indian staff were not interested in things which motivated UK developers, such as 'fast things to keep you amused' (cars, video games, hi-fi, etc.), informality and after-work drinking. Mitra had believed that the impact of Western movies, television and travel would have led to similarities in motivation structures in India, the USA and the UK. However, adopting a view of Bangalore as a 'place' as opposed to a 'space' reveals the complex social, organizational and technical networks within which actors are situated. Motivating and retaining staff requires sensitivity to these networks and social structures involving infrastructure, networks of competition from other firms, family responsibilities, attitudes to hierarchy, religion and spirituality. In one of our research reports to Sierra management we stressed that Mitra would benefit from involving himself in the local networks such as the Software Association of Bangalore, visiting colleges and giving talks at local universities. Mitra had some success with this strategy and managed to recruit a small number of graduates from the prestigious IITs, who in turn helped to attract their friends and classmates. However, heavy investments of time and money are required for implementing such a strategy. Small companies such as Sierra typically do not have such resources and Mitra, despite being the centre manager, had also to engage personally in the recruitment effort.

Knowledge workers have a strong sense of identity grounded in structures of their society, the ability to do high-tech work in a global environment and the academic community to which they belong. Schemes for recruitment, motivation and retention need to keep in view these complex processes of identity and their construction. ComSoft management tried consciously to develop this identity by creating a culture and image that would nurture the growth of such an identity. This identity is continuously being redefined in the turbulent global environment however: GlobTel tried to impose their own identity and frame of reference in the Indian operations through the 'global manager' initiative, paradoxically implemented by an Indian manager.

Training has clear implications for the GSA relationship, attrition and motivation. GlobTel's training efforts were aimed at producing the universal 'GlobTel Manager' who would encapsulate common frameworks of cultural and language, approaches to management and technical understanding. The standard 'template' of management was reflected in Sierra's attempts to create a 'little bit of England in India'. These standardization efforts aimed to manage attrition by means of expatriates 'filling gaps' and a standard 'template' being available for preparing new staff. However, management training is often done in a superficial manner, confined to such aspects of interpreting body language and speech – for example, an Indian making a sideways head movement means 'yes' and not 'no'. Such simplistic training helps to fuel stereotypes and prevents a deeper and more contextualized understanding of social behaviour. There are exceptions, of course, and we found one Indian company trying to develop a deeper understanding of Japanese culture: they even hired some Japanese anthropologists to give long-term training to Indian developers, including telling them about Japanese

culture and mythologies. Eron and Gowing also used training and informal linkages in-
directly to manage attrition and motivation. In the maturity stage of their relationship,
the onshore and offshore staff were brought psychologically closer using 'buddying', a
technique born out of scuba diving; an individual working onshore would be paired
with an opposite-number 'buddy' to link the offshore staff (in India) to the UK context
while retaining an informal link 'home' for staff located in the UK. These links back
home help to counter some of the feelings of loneliness and alienation a developer
may experience on-site and serve as an umbilical cord to the identity of the mother
company.

In developing a perspective and implications for attracting, retaining and motivating staff resources, five key questions that reflective managers need to consider are
summarized in box 11.2.

Box 11.2 Key questions: managing people

- What environmental factors influence the recruitment and retention of staff? What formal and informal
  networks can be created to provide a regular supply of trained people?
- How attractive do the developers find the company and the type of work that is being done?
- What structures of motivation can be put in place in the organization that are compatible with the
  socio-cultural context within which the developer is situated?
- How can training programmes be developed that help to foster deeper cultural understanding and
  transcend superficial information that fuels stereotypes?
- What strategies of training can be developed to try and create a 'hybrid manager' with the capability to
  'straddle' across the partnership?

11.4 Managing communication

IS development is communication-intensive, complex and difficult even in conditions
of co-location and proximity (e.g. in the same physical building). Gone are the days
of remote data-processing departments that would build systems around technical imperatives divorced from the reality of user needs. In the 1990s, methodologies and
approaches focused on user-centred design such as rapid application development,
SSADM and other similar methodologies that emphasized the importance of management process between various stakeholder groups. In the UK, the British Computer
Society 'hybrid manager' concept has been used as a basis for many Bachelors' and
Masters' IS programmes, emphasizing skills in teamworking, presentation and interviewing in addition to technical IS skills. This has led to a focus on new roles such as
the 'business analyst', designed to facilitate the *process of communication* between users
and technical developers and ensure that 'business benefits' of IT systems are realized.
These 'hybrid' managers are trained to view implementation as a challenge of 'hearts
and minds' and not just a technical exercise. Such a focus emphasizes interpersonal

(or 'soft') skills; the ability to manage expectations and politics, motivate, inspire and lead are viewed as being just as important as the mastery of 'hard' skills such as critical path analysis (CPA) and other project management techniques. This need for 'hybrid' skills becomes magnified in GSAs, with the additional challenges of language within multiple cross-cultural settings.

Practical management guides (such as Karolak 1998) suggest that communication in GSW can be understood using the 'sender–receiver' model of Shannon and Weaver (1964). Other theories categorize the various communication media according to their ability to facilitate shared meaning or convey information between remotely located senders and receivers (Daft and Lengal 1984). These models attempt to match information requirements and methods of communication; however, they have been subject to critique for presenting a functionalist, static and rational view of the communication process (Lee 1994; Markus 1994). We have adopted an approach that emphasizes communication in GSA as being better examined as an *interpretive process*, an accomplishment that is continuously being constructed and negotiated where conflict may be present and beliefs are not necessarily automatically shared. Functionalist views of communication, on the other hand, assume a mechanistic process where meaning is objective and universal and can be transmitted to neutral recipients. The complexities of the interpretive process are magnified in GSAs because development is done in a different place (usually offshore), mediated by various ICTs across time zones and by people with different cultural and technological backgrounds. Far from being neutral recipients of dialogue, GSA participants in the act of interpretation draw upon their understanding of the rules and resources of communication in the various cross-cultural systems of which they are members. Rules and resources are manifest and constructed in the communicative act in various forms, including the structure and content of an email message, a telephone conversation, a videoconference meeting or a verbal or non-verbal message in a co-located setting.

In the case studies we observed different strategies to manage communication in the use of ICTs; these ranged from the intensive use of ICTs (email and videoconference in the case of GlobTel) to a more conscious avoidance of it (Japan). Sierra adopted three project models and associated communication strategies. In model one, development relied on the assumption that ICTs would bridge communication between UK customers and developers in India. Model two involved a split-team model between the UK and India and in model three only projects that were modular, self-contained, stable and 'specifiable' were sent to India. These modules would then require minimal interaction between India and the UK and could be 'bolted in' to the master application at specified project stages. Difficulties in communication ranged from the relatively 'hard', technical problems (ICT delays, breakdowns) to 'soft' issues such as differences in accents, dissonance and background knowledge. In models one and two, misunderstandings of specifications were related to 'hard' technical problems such as unreliability of videoconferencing equipment. At the 'softer' end of the continuum, customer requirements were dynamic; technical knowledge evolved in the UK through situated practice could not easily be transferred to the Indians. The ICTs used to communicate did not offer the 'readiness at hand' which a face-to-face discussion or meeting around the water cooler would supply. Differences in background knowledge between Indian- and UK-based developers often contributed to misunderstandings about communicating and interpreting design specifications; such problems meant that UK-based developers did not adequately trust Indian development (we discuss 'trust' in more detail below) and often preferred to work locally themselves than engage with the communication problems.

In Gowing, the communication moved from an initial 'body-shopping' model using face-to-face co-situated developers to an onshore–offshore arrangement where communication was mediated through an onshore Indian presence. Communication consisted only of structured programming specifications that could be sent to India and code returned, using simple file transfer and clarification using emails. This 'minimizing interaction' strategy, however, prevented Gowing from moving more sophisticated work offshore. When they subsequently did adopt this strategy, Eron employed a speech therapist who, along with the Eron HR manager, performed regular communication and cultural awareness training. Moving up the value chain required more intensive communication, the use of 'straddlers' who could bridge communication between UK and India and greater rotation of staff in both directions. GlobTel's communication strategies varied with the different levels of work (1–4) carried out in India, and was based upon the amount of bandwidth that was in place, the use of expatriates and the degree of permissible on-site presence. GlobTel made extensive use of ICTs, including Intranet, videoconference, email, fax and telephone. At a higher level of work, GlobTel tried to do away with communication differences through their 'global manager' initiative. Japanese–Indian communication relied on improvisation and was shaped by, and was also responsible for shaping, business models, project management practices and social relationships.

As a result of advances in tools and increases in bandwidth, ICTs will no doubt facilitate improved communication between stakeholders. Software configuration management and Web-based enterprise development portals offer access via a single interface to unified software development and project management tools. Other tools being used include fax machines, conference calls, cellular phones, pagers and videoconferencing. Instant messaging, virtual meeting rooms, application sharing, centralized document storage management search and indexing can all be used to facilitate interaction. While these technologies may improve communication under certain conditions, they in themselves do not contain any inherent capacity to ensure good communication: possessing a good word processor does not create a good author! Recognizing and agreeing the nature of 'hard' and 'soft' communication issues, staff rotation, mixed teams, training programmes, use of 'straddlers', improvisation and minimizing unnecessary communication are tactics used in the case studies at various stages of the relationship. Rather than attempting to prescribe solutions, we suggest some questions that reflective managers need to consider, summarized in box 11.3.

> **Box 11.3 Key questions: managing communication**
>
> - How can business models be developed that are compatible with the possible communication strategies and vice versa?
> - What ICTs can be used that are compatible with the socio-cultural and physical context?
> - What are the national stereotypes that individuals have, and how can these be dispelled in favour of a deeper, cross-cultural and contextualized understanding?
> - How can the onshore–offshore mix be developed to help make communication more effective? What techniques (such as the use of 'straddlers') can be used?
> - How can more effective training programmes be developed to provide the right blend of communication skills and technical abilities?

11.5 Managing relationships

As discussed in the cases, GSA relationships can be seen to pass through an *evolutionary process towards maturity*. We have interpreted these phases of the relationship based on a multiplicity of considerations, including the kind of work being done, the investments being made in infrastructure and the level of trust and understanding that exists between the partners. GSA relationships are far from being static, and tend to develop in scale, scope and form over time. A typical phase in a life-cycle from initiation, growth to maturity is schematically depicted in figure 11.1 (based on Murray and Mahon 1993). Following initiation, where most GSAs 'start small' with non-mission-critical projects, an increased proportion of the development tasks tend to be moved offshore where the costs of development are cheaper. A key question then relates to how much and what kind of development tasks should be moved offshore and when, thus making managing the balance between offshore and on-site work an ongoing challenge. The onshore–offshore mix is not static and shifts depend on peaks and troughs in workload and the kinds of work being done. If the relationship moves to maturity, higher-value activities may be undertaken, involving new forms of joint work. Alternatively, the partners may separate amicably temporarily or permanently (like a 'divorce'). The time for evolution may be short or prolonged depending on events and circumstances, whether the relationship is a 'one-off'-type project or a long-term commitment.

As GSA relationships evolve over time, the contractual basis and organizational forms may also change. A trading relationship initiated with a number of 'one-off' projects based on time and material costing may over time evolve into an alliance or JV. JVs and other strategic alliances may subsequently end in the sale of one partner to the other. While not all GSA relationships follow this linear trajectory, it was relevant to the Sierra case; Sierra's GSA started with a trading relationship with offshore outsourcing companies and subsequently led to them opening up their own subsidiary. Gowing initiated its GSA strategy with 'body-shopping' on a time and material basis governed by a service-level agreement. The partners later moved to a new contract that covered the

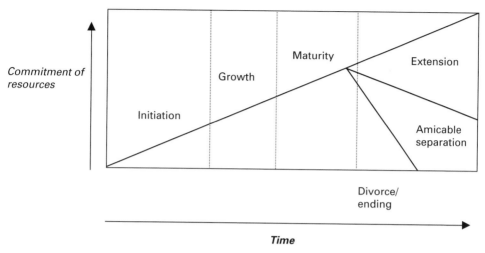

Figure 11.1 Life-cycle model of GSA relationship evolution
Source: Murray and Mahon (1993).

whole of the Cass group. In the new alliance relationship, some projects were outside of Gowing but within Cass and were valued on a turnkey or project-cost basis. In contrast, GlobTel's externalization programme started off as a 'vendor–contract relationship' and contractually remained the same after 10 years even though the content of the relationship had changed significantly over time.

The model presented in figure 11.1 does not indicate the possibility of non-linear movements in the GSA relationship, such as going in reverse, as was the case in Sierra, which had to shut down its India operations. There are various reasons for this 'non-linear' movement, including disruptive events in the industry and shifts in the balance of power and control between the partners and the consequent implications. As firms progress up the 'trust curve' to higher-value activity, the customer can become over-reliant on the outsourcer, creating a dependency situation that can be exploited through demands or refusals of price increases. At one point Gowing management carefully considered the reliance it had on Eron, and evaluated the option of appointing a portfolio of software suppliers (instead of just one) within a network of relationships that would also include Eron. Gowing eventually decided against this option because of the risks associated with new relationships, and instead entered into a closer alliance with Eron. GlobTel also considered and subsequently opened their own software development centre in India to guard against the possibilities of over-reliance on Indian firms and the consequences it might have on their activities in India.

Managing relationship evolution in GSAs is a key challenge, and one that is shaped significantly by issues of trust and its dialectical relation with power and control. Control and coordination of activity across time and space is a fundamental issue related to trust. Trust and control are not 'either–or' conditions, one presupposes and depends on the other. While control is crucial to successful completion of IS projects using the optimum resources, the need for control can be minimized in a trusting environment. Control in GSAs is concerned with the process by which one partner influences the behaviour of another and of the alliance itself (Faulkner 1999). Control mechanisms include traditional forms of equity ownership and board membership to mediate knowledge-transfer processes; training and personnel development; contracts with specified service-level agreements; and standardization of skills and processes. These *structural controls* are enforced through mechanisms such as rewards and punishments; penalty clauses; reporting procedures; client involvement in developing plans; and specifying methodologies for measuring outputs.

The relationship between trust and control can be examined from four perspectives:
- Knowledge
- Identification
- Performance
- Calculus-based trust (Faulkner and Child 1998; Sabherwal 1999).

While *knowledge-based trust* concerns shared experiences and personal relationships, *identification-based trust* is concerned with the extent of synchronizing of goals, objectives and expectations between both sides of the relationship. *Performance-based trust* is concerned with movement up the 'trust curve' through early project successes, interim deliverables, demonstrations and pilots. Calculus-based trust is related to the deterrence and rewards which are crucial to the successful completion of IS projects using optimum resources. We discuss in the following paragraphs the dialectical relation between trust and power and control.

Knowledge-based trust

While knowledge-based trust is concerned with shared experience and personal relationships, it is 'grounded in the other's predictability – knowing the other sufficiently well so that the other's behaviour is anticipatable' (Lewicki and Bunker 1996: 121). At Sierra, developers in the UK lacked the assurance of knowing their Indian partners sufficiently enough so that their 'behaviour was anticipatable'. The security and comfort offered by a well-understood partner was undermined by communication problems and 'misinterpretations', caused in part by differences in respective background knowledge and worsened by the problem of attrition. The lack of informal socialization in a 'community of practice' contributed to both sides not being able to understand the thinking or reasonably predict the actions of their partners in India or the UK. Indian companies are attempting to manage knowledge-based trust better by localization of their offices in North America or Europe to help present a local 'front-end' for direct liaison to enable more effective knowledge transfer. This 'front-end' is enabled through the opening of subsidiaries, acquiring or partnering with foreign companies to gain access to local domain experience and markets. Similarly, North American and European software houses have themselves opened subsidiary development centres or partnered with Indian companies to offer their customers an 'indirect' route to out-sourcing through a local provider. Sierra attempted this route; another example is Arrk (www.arrk.net) which has an established JV with a Bombay software house. At the time of writing Arrk was planning to open their own Bombay subsidiary.

The subsidiary arrangements of the major companies such as Texas Instruments have contributed to knowledge of the offshore development process. Various middlemen, often experienced, well-connected ex-employees can help facilitate and foster relation-ships or set up and incubate offshore subsidiaries. Government and quasi-governmental organizations are tasked with facilitating GSA. The UK Department of Trade and In-dustry's 'Enterprise India Initiative' (www.tradepartners.gov.uk) claims considerably to smooth the GSA maturity process. The DTI provides a partner-matching service, expert advice and organized trips to visit India and suppliers. For some of these reasons, GlobTel proposed setting up it offshore development centre in India to deal with the knowledge gaps it felt existed in its understanding of India – for example, how to deal with attrition. This move to come closer to India was, however, seen by the Indian firms as an exercise of power and control, leading to tighter micro-management. MCI's initiative to develop a proximity development centre in North America, with a view to developing a closer understanding of GlobTel's management practices – that remained largely tacit despite the many years of the relationship – was rejected by GlobTel on the grounds that it would add to development costs. We argue that while knowledge-based trust is useful in developing shared knowledge and expectations, it has continuously to contend with the implications for power and control.

Identification-based trust

Identification-based trust is concerned with the synchronizing of goals, objectives and expectations between both sides of the relationship and operates in different ways. At the strategic level as well as the operational levels of the relationship, identification is linked closely to the role of 'champions'. Gowing and Eron's relationship was founded on the strong personal relationship and friendship between the Eron Managing Direc-tor based in the UK and Jones, Gowing's Managing Director. Jones became a powerful GSA 'champion' and strongly advocated Eron moving up the value chain partly through outsourcing and its extension to other parts of Cass. By virtue of their Indian roots, Ghosh and Paul similarly acted as 'shock absorbers' in the GlobTel–India relationships. They served as conduits to develop local knowledge and to deal with crisis situations as and when they developed. However, both the Indian and North American management often criticized this strong identification as it was felt to impede developing stronger business oriented control structures. In Sierra, Mitra was seen to have a similar 'cham-pion role' on account of his Indian origins and the identification that this implied. However, Mitra's understanding of the complexities in India was rather superficial and was reflected in his attempts to develop strong structures for power and control (to deal with the issue of 'leakage', for example). The adverse effects might have been mitigated by development of a stronger sense of identification with his Indian colleagues. In Sierra, at the project level, developers at the two sites did not share the close 'bonding friendships' crucial for this type of trust to evolve. The physical and cultural distance between the sites made informal socialization impossible to achieve, and the staff could not genuinely 'act for each other'. In contrast, at the mature stage of the relationship the Gowing and Eron staff shared a higher level of identification with each other's needs, and this led to close bonding friendships.

Performance-based trust

Performance-based trust concerns building up confidence with respect to the ability of the partners to effectively carry out the tasks they are supposed to on time and on budget. In the GlobTel case, there was initially limited performance-based trust as the Indians had little prior experience in telecommunications and also in the conduct of GSW. But they had the willingness, aptitude and intellectual capacity to learn, and in the process they provided GlobTel with the potential for such performance-based trust to develop in the future. Efforts to build this performance-based trust had to confront a number of tensions along the way, for example the fact that telecommunications and software presented very different domains of expertise. These differences had serious implications for the relationship, such as in dealing with the question of attrition and in the implementation of and compliance with various control structures. As ComSoft gained in performance-based trust and obtained more ownership of GlobTel technology, it developed the confidence and need to forge its own growth path by exploring new Japanese markets. This led to a shifting of the power balance and a lessening of dependence on GlobTel for survival and growth. Sierra's ambitious goals and early failures with performance (reflected in the issue of 'leakage') led to an organization-wide dissonance. This contrasted with the Gowing case, where Eron quickly fulfilled the aspirations of Gowing management and moved along the GSA life-cycle and up the value chain. As a result of this performance-based trust, the Eron developers were used to create a 'rock of discipline' within the UK framework, attempts resisted by some groups in the Cass group (ex-RDC people) who felt that their own performance was being undermined.

Calculative trust

Calculative trust is related to deterrence and the rewards crucial to successful completion of IS projects using the optimum resources. This form of trust tends to be strongly associated with the early stages of GSA relationship growth, before other forms of trust (performance and identification) grow. Calculative trust reflects a dialectical relationship that can be metaphorically viewed as 'walking a tightrope'. Project teams, for example, walk a tightrope between an exercise of freedom to organize their work and deliverables, while being subjected to exhaustive reporting and micro-management controls. At Gowing, in the early stages of the relationship, structural controls dominated through the use of structured methodologies and extensive reporting tools, including key performance indicators (KPIs) represented on a 'dashboard' on the Intranet. Strategic-level controls were exercised through varying the size of contracts and leaving open the potential for further contracts. During the later stages, with growing maturity and Eron's increasing integration within Gowing, the relationship depended more on knowledge and identity forms of trust. At GlobTel, standardization was a key force for long-distance control of Witech, supplemented through the use of expatriates and training programmes within the 'global manager' 'template'. Similar attempts at standardization in ComSoft met with resistance – for example, their refusal to have an expatriate located on their premises. The ComSoft and Witech cases provide contrasting examples of how GSA strategies tread a thin line between control and standardization on the one hand and the need to nurture local identity and creativity on the other hand.

In summary then, the lens of a dialectic between control and trust through the different stages of a GSA relationship is useful for managers to develop the implications of managing the evolving process. This lens helps to formulate some key questions, summarized in box 11.4.

Box 11.4 Key questions: managing GSA relationships

- What is the role of GSA 'champions' in helping achieve the objectives of the relationship?
- How can various forms of trust be developed and integrated for effective performance and the growth of maturity in the relationship?
- What control mechanisms are appropriate at different stages of the relationship and how do they impinge upon issues of trust? How should control be balanced against such different forms of trust ?

11.6 Managing ethically

It is important that any book concerned with the globalization of work should address ethical concerns and questions. Standards for global ethical practice have been proposed by three main groups:
- Private sector
- Public sector
- Societal groups.

Private sector standards include ISO which focuses on internal management systems, addressing such topics as conflicts of interest, disclosure and employee conduct. Codes of ethical practice also exist for the IS and computing profession such as the ACM, IEEE and British Computer Society (BCS) (Spinello 1995; Johnson 2000). Public sector standards at a high level are proposed by UN (www.unglobalcompact.org) and OECD (www.oecd.org) guidelines for multinationals. *Societal and non-governmental organizations* (NGOs) such as research and academic institutions focus on the harmful effects of globalization. A popular text critical of the impact of globalization is Klein (2000), who outlines many of the concerns of societal and NGO groups, often categorized as 'anti-globalization'. Klein and other authors (such as Korten 1995; Faux and Mishel 2000; Shiva 2001) critique the power of MNCs with operations in developing countries and the inequality and ecological catastrophe associated with unbridled global capitalism. Shiva (2001) asserts that globalization is not merely a geographic phenomenon that is tearing down national barriers to capital but also tearing down ethical and ecological limits on commerce. A specific concern of Klein is the 'sweatshop' conditions and exploitation of cheap, sometimes bonded, labour in developing countries. Shiva (2001) draws attention to environmental issues and the export of dirty, polluting industries to 'places where life is cheaper'. This point is also echoed by Beck (1992) in his discussion on the 'redistribution of risk', often starkly exposed through cases such as the Union Carbide chemical disaster in Bhopal in 1986.

Dealing with such risks raised by the 'anti-globalization' groups requires a serious consideration of ethics in developing management action. The criticisms of the 'anti-globalization' groups against MNCs are important, even if viewed only from a pragmatic perspective of not tarnishing the corporate image. The work of NGOs such as Greenpeace has had adverse public relations (PR) consequences for MNCs, as reflected in examples of direct action such as burning genetically modified (GM) crops and organized protests at Starbuck's coffee shops and Disney toy stores. Such protests are not good for recruitment, share price or investor confidence.

However, no complete code of ethics currently exists for the conduct of GSW. In the discussion to follow we therefore argue the important need to establish such a code, drawing on the standards of the three groups outlined above and laying down the tenets of what issues such a code of ethics should address.

The international GSW market is generally not prone to the worst excesses of 'sweatshop' conditions or bonded or child labour. At the time of writing a global IT skills' shortage has helped further to legitimize outsourcing strategies. In GSAs, managers need to deal with concerns over PR image arising from widespread unease about job losses and immigration; these issues have become particularly pertinent following the 11 September attacks in New York. At the lower end of data-processing outsourcing, low pay, high surveillance, and high pressure may raise ethical concerns similar to those raised by other forms of global work. Although difficult to characterize as 'sweatshops', some outsourcing software houses may work on precarious employment contracts, Taylorized work processes in software factories without the support of union organization and poor attention to health and safety. The Taylorized form of work and the principles of switching of resources, on which it is based, render work mobile that can be moved without much fuss to cheaper production centres (Mowshowitz 1994; Pearson and Mitter 1993; Korten 1995). In the emerging GSW supplier market, the consequence may be a spiralling 'race to the bottom' in which companies try to cut each other's throat on prices in order to gain international contracts. The consequences of this price-cutting could lead to declining labour standards, poor wages and social conditions tending towards 'worst' rather than 'best' practice.

Current developments in offshore call centres are also a cause for ethical concern. Call-centre workers may be asked to introduce themselves with Western names, neutralize their accent and work nights under high-pressure conditions with extensive surveillance and limited union protection. Implications of the movement of this kind of work to the cheapest possible locations and the effects on job losses are not yet fully understood. The ethics of a cosmetic subterfuge of accents, Western names and the range of activities that offshore call centre operatives perform need to be actively debated in a framework that balances economic considerations with broader societal impacts. Such debates could lead to the formulation of sensitive regulatory mechanisms which also need to take into account that call centre-work does not need to be static, and could lead to other kinds of future activities. A process perspective that has been emphasized in our analysis permits us to visualize this future potential. For example, while medical transcription work can currently be criticized for Taylorization, the future might open up the possibility of higher-end telemedicine-based activities that could draw upon the potential of a highly competent source of Indian medical doctors to provide expert consultations.

At a policy level, there are significant undercurrents of unrest about the activities of MNCs. A column in the *Times of India* (Dubashi 2002) argued that under the cloak of globalization and liberalization, power was being transferred to foreign business led by MNCs. Indian policy has a history of suspicion towards MNC power and companies like IBM and Coca-Cola were asked to leave in the 1970s. While the government through its liberalization policies in the early 1990s acknowledged the simplistic nature and adverse effects of previous policies, the culture of suspicion is not yet buried: MNCs are still viewed by some, especially within government circles, as representing a new wave of colonialism that will undermine the interests of the poor. Such MNCs are seen to have as a primary objective foreign shareholder satisfaction and maximization of profits. Hutton and Giddens (2001) discuss the concentrations of transnational power and call for a reworking of the role of the WTO as a power that 'underwrites and polices a basis framework of rules of capitalist engagement rather than enshrine free trade as an absolutist principle' (2001: 220). State and national governments have an important role to play in ensuring that MNCs operate in an ethical manner. Progressive states such as Andhra Pradesh are playing a lead role in this regard by attempting to draw up adequate regulation through legislative acts, while at the same time providing conditions in which MNCs can operate in an enabling environment. How this balance between regulation and enablement is met in practice is an open empirical question, and it should be unhesitatingly recognized that governments must play a lead role.

It is unclear what the future role of the WTO will be and whether it will be able effectively to act as 'global policemen' protecting against the worst excesses of globalization.

There is clearly a need for managers involved in GSA in various relationship forms, on both sides of the relationship, seriously to consider issues of corporate and social responsibility. This should involve a full and explicit consideration of the economic, social and human rights of various stakeholders, that transcends not just the shareholders' interests but also includes the concerns of society at large. The code of ethics should address the various accountability concerns raised by NGOs, governments, citizens and employment law.

The Gowing case provides some illustrations of failure in corporate responsibility. The GSA was used to facilitate change and exert power and control over UK employees; this subsequently led to the layoff of existing staff, avoiding the stipulations of UK employment law. Both Gowing and Eron failed to consider issues beyond the pure economics of the situation and actions of management can be seen as an illustration of the globalization critique where the power of large organizations overrides the concerns of workers in the pursuit of increased profits. The case paints a picture of globalization as a relentless economic machine. From the perspective of anti-globalists, there is a need to protect the local from organizational actions that can potentially lead to adverse societal effects.

Ethics is not an exact science, and what is *lawful* may not necessarily be *ethical*. Starting points for promoting an ethical approach to GSA need to consider the corporate and social responsibilities and obligations to various stakeholders (Spinello 1995). In this context, De George's (2001) seven moral guidelines for MNCs can be seen as instructive for those setting up GSA-based subsidiaries:
1 Do no intentional direct harm
2 Produce more good than bad for the host country
3 Contribute by their activities to the host country's development
4 Respect the human rights of employees
5 Pay their fair share of taxes
6 Respect the local culture and work with it not against it
7 Cooperate with local government in the development and enforcement of just background institutions for tax, health and safety, etc.

The basic moral norms in points 1–3 suggest that MNCs will do good only if they help the host country more than the harm it and avoid exploitation. Point 4 relates to setting minimum standards for pay as well as health and safety issues. Trends in 'reverse globalization' (as emphasized in the Gowing case) imply that the rights of employees in developed countries, through job losses for example, must be considered on a par with those in developing countries. MNCs inevitably produce some changes in the cultures in which they operate and this can be advantageous through the spread of best practices. However, rather than simply transplanting global or North American or UK ways of working, MNCs need to adopt an ethical framework that takes into account local needs and customs; such a framework could go a long way towards hybridized models of work that aim to celebrate rather than suppress diversity.

In outsourcing arrangements, building hybridized models is easier said than done, as companies on both sides need to emphasize stakeholder interests while dealing with conditions of extreme organizational complexity and turbulence. Despite these contextual conditions, serious attempts need to be made to consider the basic moral principle of doing no intentional harm. From a practical perspective, ethical issues can be emphasized by categorizing them into concerns of IP, privacy and surveillance/security:

- IP is an important issue in GSA that in developed countries is legally enforced through trade secrets, patents and copyright laws. Many countries take a different legal and ethical approach to IP: the legal position in Korea and Japan considers that ideas should be shared freely for the good of society since knowledge is communal and belongs to the public domain.
- Privacy and surveillance/security issues are of importance at the employee level. The methodologies used at Gowing to reflect transparency in project activities and the bug-fixing systems at GlobTel are examples. Gowing's system of developing transparency involved displaying on the Intranet the name of the individual assigned the task of fixing the bug and the time taken to do it. Building transparency in one setting may, however, infringe laws in another; in Germany, it is not legal publicly to display the name of an individual. Other issues concern the privacy of confidential information that may be passed between companies and into the environment, with potentially deleterious consequences. Security and protection of customer data from hacking, and deliberate damage of the software code through viruses or otherwise may be particularly pertinent in the event of staff attrition and the moral responsibilities an individual should display.

As well as pressure from NGOs and consumer groups, unions are also voicing concerns about the implications of GSA for local job losses. We witnessed these concerns at a workshop that we organized in Berlin on the topic of international IT skill shortages. If companies do not self-regulate, employment law will be needed to deal with various forms of offshore outsourcing. Articulating a GSA oriented ethical policy with respect to job losses is a broader responsibility of software industry associations in both onshore and offshore locations. The role of government is also important to prevent the 'race to the bottom' through undercutting of prices by competing software firms both within the country and globally. Indian firms are competing with each other and also the new and lower-cost locations of China and Vietnam, for example. While MNCs face competitive pressures for survival, there is an important role for governments to ensure that the competitive basis of firms' actions is not attained at the cost of the lowest ethical position. Most importantly, policy responses from the GSA participants themselves in a statement of corporate responsibility can be included in outsourcing contracts within the framework of a service-level agreement.

In developing a perspective for managing ethically and appreciating implications, there are five key questions that need to be considered, summarized in box 11.5.

> **Box 11.5 Key questions: managing ethically**
>
> - What is the firm policy on GSA corporate responsibility? How can it be articulated and disseminated?
> - How can firm policy on ethical and social responsibility be realized practically within both partner organizations?
> - Who are the 'visible' and 'invisible' stakeholders? What are the different impacts (broader than economic) of the GSA on these groups?
> - What is the policy framework in the host countries (related to anti-piracy, for example) to recognize and protect ethical concerns?
> - What is the role and power of industry associations to advocate and legislate ethical concerns?

11.7 Conclusions

This chapter has summarized the implications for managers around the six analytical themes discussed in the book. We have addressed these implications by discussing the relevance of the broad themes of the 'out-of-control' nature of global operations, the importance of adopting a network perspective and the complex nature of risk assessment and management. Through our analysis, we developed implications in the form of a set of key questions that need to be addressed in GSA management. These questions related to managing knowledge, people, communication and relationships within a broader ethical framework. The focus has not been on suggesting recipes and frameworks for successful GSA management, but on providing conceptual 'templates' within which learning and thinking about GSA processes can occur. Managers can draw lessons from these cases along with other literature and augment their own experience. What we present are in-depth accounts of how particular companies have managed their GSAs, and their stories which we have interpreted and theorized. These experiments, we hope, will be of benefit to the critical, reflective manager.

BIBLIOGRAPHY

Apte, U. (1990). Global outsourcing of information systems and processing services, *The Information Society*, 7, 287–303

Apte, U. and Mason, R. (1995). Global disaggregation of information intensive services, *Management Science*, 41, 7, 1250–62

Beck, U. (1992). *Risk Society: Towards a New Modernity*, London: Sage

Carmel, E. (1999). *Global Software Teams: Collaborating Across Borders and Time Zones*, Upper Saddle River, NJ: Prentice-Hall

Castells, M. (1996). *The Rise of the Network Society*, Oxford: Blackwell

Daft, R. and Lengel, R. (1984). Information richness: a new approach to managerial information processing and organization design, in B. Staw and L. Cummings (eds.), *Research in Organizational Behaviour*, 6th edn., Westport, CT: JAI Press, 191–233

De George, R. (2001). Ethical dilemmas for multinational enterprise: a philosophical overview, in A. Malachowski (ed.), *Business Ethics: Critical Perspectives on Business and Management, vol. III, International and Environmental Business Ethics*, New York: Routledge

Dubashi, J. (2002). Into foreign hands: India's reverse transfer of power, *The Times of India*, 17 January, 14

Faulkner, D. (1999). Survey – mastering strategy: trust and control in strategic alliances, *Financial Times*, 29 November

Faulkner, J. and Child, D. (1998). *Strategies of Co-Operation: Managing Alliances, Networks and Joint Ventures*, Oxford: Oxford University Press

Faux, J. and Mishel, L. (2001). Inequality and the global economy, in W. Hutton and A. Giddens (eds.), *On the Edge*, London: Vintage, 93–111

Giddens, A. (1990). *The Consequences of Modernity*, Cambridge: Polity Press

Hutton, W. and Giddens, A. (2001). *On the Edge*, London: Vintage

Johnson, D. (2000). *Computer Ethics*, Englewood Cliffs, NJ: Prentice-Hall

Karolak, D. (1998). *Global Software Development: Managing Virtual Teams and Environments*, Los Almitos: IEEE

Klein, N. (2000). *No Logo*, London: Picador

Korten, D. (1995). *When Corporations Rule the World*, London: Earthscan

Lee, A. (1994). Electronic mail as a medium for rich communication: an empirical investigation using hermeneutic interpretation, *MIS Quarterly*, 18, 2, 143–57

Lewicki, R. and Bunker, B. (1996). Developing and maintaining trust in work relationships, in R. Kramer and T. Tyler (eds.), *Trust in Organisations: Frontiers of Theory and Research*, Thousand Oaks, CA: Sage

Markus. L, (1994). Electronic mail as the medium of managerial choice, *Organisation Science*, 5, 4, 502–27

McFarlan, F. (1995). Issues in global outsourcing, in P. Palvia *et al.* (eds.), *Global Information Technology and Systems Management: Key Issues and Trends*, Nashua, NH: Ivy League Publishing

Mowshowitz, A. (1994). Virtual organisation: a vision of management in the information age, *The Information Society*, 10, 267–88

Murray, E. and Mahon. J. (1993). Strategic alliances: gateway to the New Europe?, *Long Range Planning*, 26, 102–11

Pearson, R. and Mitter, S. (1993). Computerisation, work and less developed countries, in M. Ermann, M. Williams and M. Shauf (eds.), *Computers, Ethics and Society*, Oxford: Oxford University Press

Rajkumar, T. and Mani, R. (2001). Offshore software development: the view from Indian suppliers, *Information Systems Management*, Spring 63–72

Sabherwal, R. (1999). The role of trust in outsourced IS development projects, *Communications of the ACM*, 42, 2, 80–6

Shannon, C. and Weaver, W. (1964). *The Mathematical Theory of Communication*, Urbana: University of Illinois Press

Shiva, V. (2001). The world on the edge, in W. Hutton and A. Giddens (eds.), *On the Edge*, London: Vintage, 112–29

Spinello, R. (1995*). Ethical Aspects of Information Technology*. Englewood Cliffs, NJ: Prentice-Hall

Index